SMART LOGISTICS
& SUPPLY CHAIN
INFORMATION PLATFORMS

www.royalcollins.com

SMART LOGISTICS & SUPPLY CHAIN INFORMATION PLATFORMS

WANG XIFU & CUI ZHONGFU

Books Beyond Boundaries

ROYAL COLLINS

Smart Logistics & Supply Chain Information Platforms

Wang Xifu & Cui Zhongfu
Translated by Daniel McRyan

First published in 2022 by Royal Collins Publishing Group Inc.
Groupe Publication Royal Collins Inc.
BKM Royalcollins Publishers Private Limited

Headquarters: 550-555 boul. René-Lévesque O Montréal (Québec) H2Z1B1 Canada
India office: 805 Hemkunt House, 8th Floor, Rajendra Place, New Delhi 110 008

ISBN: 978-1-4878-0906-5

To find out more about our publications, please visit www.royalcollins.com.

FOREWORD

China is in the phase of transformative adjustment and upgrade in the economy, guided by innovation and reform. Taking full advantage of the Belt Road Initiative, a broader opening-up to the world is happening. The guiding policies behind this were proposed as part of the *Made in China 2025* strategy.

This strategy is driven by innovation; a focus on quality; green development; optimized structure; and should be talent centered. Basic market principles take a leading role, with government guidance on the present and long-term prospects of the industry and its general trajectory, while independent development and open cooperation are also emphasised. China's modern logistics industry is becoming more automized, data-centric and smart, thereby improving overall efficiency and reducing costs.

Over the past few years, smart technology – including artificial intelligence, the Internet of Things, blockchain, and big data – have been preliminarily applied in logistics. This intelligent hardware has been highly effective and led to a new business form: ultimately the entire process has become 'smart'.

Smart logistics takes 'Internet +' as the core, and relies on the technical support of cloud computing, the Internet of Things, mobile Internet, big data, artificial intelligence and blockchain. The seamless integration of information systems in all links of the logistics network can be tracked, sensed and monitored in real-time. This is a major difference compared with the older system.

For the logistics and supply chain industries, smart logistics embodies the value of data, of connection, and of collaboration. Data helps logistics companies grasp user needs and enhance corporate efficiency and brand image. Connection refers to the real-time synchronization of operation and logistics information that connects the internal and external elements of a logistics company to achieve intelligent management and information sharing; online all-elements of logistics promotes the overall interconnection of the supply chain and leads new smart logistics development. The value of collaboration is reflected in the promotion of collaboration and symbiosis between logistics companies and other industries and between supply chain node enterprises to realize the collaborative management of the supply chain; the mutual benefits of the logistics system, the financial system, the marketing system, and the data service system can generate huge synergy and help build a complete system; with the integration of the Cloud and the Internet, a new trend of industrial development comes into being where the Internet, cloud storage and cloud platform collaborate.

According to the positioning and development requirement of modern logistics, business clusters are formed, including integrated logistics services, bulk commodity supply chain services, bulk commodity trading services and logistics park integrated services; system integration technology is applied to comprehensively integrate modern logistics and supply chain information systems throughout the entire process; a smart logistics information platform, a bulk commodity supply chain information service platform, a bulk commodity electronic trading platform, and a logistics park information service platform are built to achieve lower costs, greater efficiency, and higher management and service levels. Consequently, a smart logistics and supply chain information platform with integrated resources and advanced technology, serving all regions of the country, comes into being.

Following the third industrial revolution, 2013 was the first year of big data, 2014 of mobile Internet, and 2017 of artificial intelligence. In this context, the core idea of this book is the deep and comprehensive integration of intelligent technologies and the logistics industry. Oriented at the application of mainstream information technology in modern logistics, it concentrates on the research of intelligent logistics information platform systems, theories, methods and technical applications in the new age, and promotes the healthy, orderly, and coordinated development of Chinese logistics.

This book promotes the intensive development of logistics, and offers a platform for people to learn and master new ideas and technologies in the face of the rapid development of artificial intelligence, big data, Internet of Things and cloud computing. The book has a complete system, clear arrangement of ideas, detailed technology, accurate data, and approachability, thus it can promote in-depth development of smart logistics to a higher level and serve as a decision-making reference and work guide for government departments at all levels, logistics enterprises, and the information industry.

To write this book, we have consulted extensive scientific and technical literature. And we would like to express our heartfelt thanks to the authors of the relevant literature. In the meantime, having conducted multiple surveys on relevant logistics companies, we were able to integrate the ideas of many technical personnel and experts, to whom we would like to express my most sincere gratitude. In addition, Zhang Wenying, Dai Lufeng, Guo Zhouxiang, Liu Haibin, Sang Junjing, Zhang Hao, Long Yingting, Ma Junchi, Yang Liu, and Jiang Jiake have also contributed to the writing.

Due to the limited personal capability and time, the rapid development of modern logistics technology and the logistics industry, and the continuous upgrade of related technology and management concepts, some mistakes cannot be avoided, thus we are open to criticism and correction.

October, 2018

CONTENTS

CHAPTER 1

BACKGROUND AND DEMAND ANALYSIS OF PLATFORM CONSTRUCTION

1.1 Background

The guiding policies of *Made in China 2025* proposed that innovation should act as the driver; quality should come first; green development should be encouraged; structure should be optimized, and talent should take center stage. Also, market principles should play a leading role with government guidance; a focus on the present with heed to long-term prospects; key breakthroughs; independent development, and open cooperation.

China's modern logistics is developing in the direction of automation, intellectualization, and datamation, thereby enhancing overall industrial efficiency while cutting costs. In the past few years, intelligent technologies such as artificial intelligence, the Internet of Things (IoT), blockchain, and big data have been predominantly applied in logistics. The application of smart technology is a chief component of rapid development in the industry, leading to a new form of business logistics. Consequently, operating costs are reduced and profits are increased.

Smart logistics takes Internet + as the core and relies on the technical support of cloud computing, IoT, mobile Internet, big data, artificial intelligence and blockchain. The seamless integration of information systems at all levels of the logistics chain creates a structure that can be identified, traced and monitored automatically in real-time. This precipitates immediate response and smart decision making.

Smart logistics integrates a variety of service functions and embodies the characteristics of modern economic operation. Focused on the fast, efficient and smooth process of information, business, and fund flow, it is an effective propellent for the logistics industry and supply chain.

This integrated development between smart logistics and the supply chain establishes a platform where multiple parties – including manufacturers, wholesalers, retailers, and carriers – collaborate, share information and resources. This enhances resource utilization and productivity, helps achieve business connections among supply chain members, attracts capital, develops value-added services, builds healthy development, and creates richer social values. Consequently, a smart logistics and supply chain information platform with integrated resources and advanced technology, serving all regions of the country, comes into being.

1.2 Construction goals and positioning

1.2.1 Goals

Pivoted on the construction of a system operation and program analysis, the overall objective of the logistics and supply chain information platform are summarized as follows:

1. A strategic platform oriented to all fronts of the logistics industry
The construction of a smart logistics and supply chain information platform enables information and resource sharing among business operations, transportation, warehousing, distribution, and expansion of logistics finance. It has also allowed full-range supervision of the industry and helps logistics companies to carry out supply chain innovation, transformation, and upgrade. Ultimately, it provides information support for the integrated management of intelligent logistics and supply chain services. In turn, this safeguards the entire smart operation strategy of the industry.

2. A cooperative platform for smart logistics and full supply chain management
The smart information platform collects production, transportation, and sales information of related enterprise supply chains engaged in the logistics business. This helps realize the sharing of information resources throughout the supply chain; thus, all links, members, and enterprises can fully control their own logistics business, exchange information, monitor the supply chain, and intelligently manage the whole process between upstream and downstream related enterprises in the supply chain.

3. A development platform for a smart supply chain based on resource integration
The smart supply chain is an important topic regarding supply chain innovation. For example, how can one apply artificial intelligence to optimize their inventory, or how to realize automatic replenishment and pricing for retailers? The establishment of this platform can help integrate information between the design and business sectors of logistics. With logistics and supply chain information linked, the extensive sharing of resource information between various businesses and sectors is possible. This helps accelerate innovation and development.

4. An information platform integrating logistics with the supply chain

Fine communication infrastructure is built to enable object-oriented data exchange based on the functions of the four business clusters: modern logistics; bulk commodity supply chain service business; bulk commodity trading business; and the logistics park service business. This fully utilizes the four corresponding sub-platforms: smart logistics information; bulk commodity supply chain information service; bulk commodity electronic trading, and the logistics park information service. Consequently, it achieves the informatization and intelligent management of upstream and downstream enterprises in the supply chain and improves the utilization rate of infrastructure.

5. A technical platform based on key technologies such as cloud computing, IoT, mobile Internet, big data, and artificial intelligence

The system integrated technology is able to make the sub-platform and its subsystems an integrated and more powerful new platform. With the help of cloud computing, IoT, mobile Internet, big data, and artificial intelligence, it attracts various industries and enterprises, building their information systems on integrated platforms. This meets the continuous and growing demands of all parties in the supply chain for smart logistics and supply chain information.

1.2.2 Positioning

Based on its overall construction and combined with modern logistics industry clusters, the smart information platform, in line with the national and regional guiding policies, is positioned as follows:

1. A global platform that leads industry and guides enterprise

From a macro perspective, smart logistics can supervise, track, and guide the entire process from all angles. The macro-control via smart logistics and the supply chain information platform is more transparent, time-saving, efficient, and simplified. From the government to the industry and to enterprise – the platform enables a full supply chain to achieve comprehensive, intelligent and practical macro-monitoring. It is a global platform.

2. An extended and innovative platform centered on service

The smart logistics and supply chain information platform not only renders a full range of information services to suppliers, carriers, retailers, and customers, but also, with the help of business and management model thinking innovation, combines supply chain links with new types of logistics, finance, and sales. This extends the scope and duration of services to customers with value-added services in the chain. It is an innovative platform.

3. A reliable platform for information sharing and secure intercommunication

The platform integrates information in all aspects of smart warehousing, transportation, loading and unloading, handling, packaging, distribution, supply chain, etc. With safety as the priority, it enables information sharing of all members as well as visualization and traceability of logistics. Under the premise of ensured security of supply chain information, it plays the role of a reliable, interconnected and efficient bridge of sharing; it is a reliable platform.

4. A technical and practical platform for linkage and sharing of resources
Based on a variety of technologies, the identification and control of key links in the supply chain realizes the sharing of resources, builds a multi-technology-run structure and creates a huge resource pool. It meets the growth demands of applications and customers with technicality and practicability; it is a practical platform.

5. A smart and integrated platform for overall planning and strategy
The platform connects all levels and aspects of logistics, links separate business and information flow, and procurement, transportation, warehousing, and distribution – forming a complete supply chain. It creates a collaborative working mechanism with the help of information sharing to ensure that the supply chain is accurate, timely, efficient, and unobstructed.

By harnessing modern network economic development laws such as resource services and service customization, it smartly provides customers with accurate business in the supply chain, forming a complete coordinated closed loop; it is a smart and integrated platform.

1.3 Demand analysis

1.3.1 Development status of the logistics industry

The new normal of Chinese logistics, which is at a critical point of transformation and upgrading, has presented arduous challenges as well as strategic opportunities for the development of smart logistics and supply chains. In light of the new situation, the Chinese logistics industry will concentrate on quality and efficiency, seek strategic breakthroughs, cultivate new competitive advantages, comprehensively upgrade, and respond to the new normal of logistics. All actions are reflected in the development of the four new concepts.

As a major manufacturing power, China will be challenged with raw material procurement, productivity layout, and product marketing on a global scale, as the need to promote Chinese manufacturing overseas becomes increasingly urgent. Therefore, companies must strengthen the layout of key logistics nodes and their control over logistic resources. This will integrate supply chain management, establish a global supply chain system, realize the global allocation of resources, help build a cooperative partnership with global stakeholders, and dominate the supply chain.

Taking the linkage of supply chain platforms as a breakthrough point, industrial logistics has been vigorously developed; taking the integration of supply chain platforms as a breakthrough point, this network has gradually been perfected – eventually a complete all-in-one supply chain information platform is created.

In 2009, International Business Machine (IBM) proposed building a future-oriented supply chain with the three major characteristics of advancement, interconnection and intelligence. Sensors, radio frequency identification (RFID), actuators, global positioning systems (GPS) and other equipment would be used.

Next, it inspired the concept of smart logistics. Unlike the construction of a virtual Internet management system, smart logistics pays more heed to integrating the IoT and sensor networks with the current Internet. Supported by scientific management, it strives to achieve automation, visualization, controllability, intelligence, and networking, thereby improving resource utilization and productivity, and creating more social value. Considering the array of benefits, it is of particular importance to build a complete smart logistics information platform and implement this in the logistics industry.

The development of modern logistics is impossible without the exchange of information. Accordingly, this inspires the concept of the sharing economy – where offline idle goods or services are integrated so they can be sold at a lower price. The development of the sharing economy is a process of disintermediation and re-intermediation. Its advent has broken the dependence of laborers on commercial organizations. Instead, they can directly render services or sell products to end users (effectively cutting out middlemen); although individual service providers are separated from commercial organizations, to widely reach the demand side they are connected to the sharing economy platform of the Internet.

The emergence of this platform helps individual labor solve the office space model problem through WeWork (maker space) and capital P2P loans (Internet lending platform) at the front, and addresses the difficulties of gathering customers at the back end. In the meantime, its customer gathering effect enables individual business to better focus on creating quality products or services.

In conventional logistics operations, large coal production enterprises had to manually record the delivery time, type of goods, truck information, etc. during freight operations. However, the creation of the car-free carrier platform changed this situation greatly as code scanning systems automatically identify license plate numbers and weight data. The entire process – from hiring the truck driver to successful delivery – is therefore recorded. This new model of car-free carrier relies on the Internet to build a logistics information platform. It then integrates and dispatches scattered logistics resources such as vehicles, station yards, and sources, fundamentally altering conventional manual logistics to a more automized and integrated process.

The 19th National Congress of the Communist Party of China (NCCPC) report in 2017 proposed making bold changes without ossification and stagnation. In order to adapt to the new ideas of economic transformation and development, online and offline resources in logistics are efficiently integrated to accurately and quickly complete tasks and meet customer requirements. The demand for informatization construction – namely a smart information platform centered on the four major platforms (smart logistics information, bulk commodity supply chain information service, bulk commodity electronic trading, and logistics park information service), is particularly urgent.

1.3.2 The basis for the integrated development of modern logistics and supply chains

Through analysis of the contemporary situation in logistics, corresponding information platforms are built to initiate the integrated construction of smart logistics and supply chains by business sectors. Construction status and basic analysis are presented as follows:

1. Development status of integration of smart logistics and supply chain
Inconsistencies in the info ports of information platforms has massively reduced the efficiency of data transmission for modern logistics enterprise, yet the hidden dangers of information security have increased. Smart logistics has not yet been fully implemented. Unshared data and the lack of a fully transparent record have made it difficult for integrated construction to occur. In terms of business cooperation, most entities in the Chinese supply chain harbor mistrust towards their fellow companies, thus inter-company cooperation stays mostly at the level of interest needs. In China, these obstacles to the integrated development of smart logistics directly affects companies involved and the industry at large.

2. Basic conditions for the integrated construction of modern logistics and supply chains
The construction of the smart platform is based on the logistics park, which is required to possess the fundamental conditions for platform construction. The logistics industry has an advantageous and solid industrial base, playing a leading and demonstrative role in logistics transportation, third and fourth party logistics, cold chain transportation, logistics real estate development, smart logistics, and merchandise trade. Therefore, smart supply chain concepts can be applied to various fields; there is also advanced technical equipment which contributes greatly to supply management collaboration between logistics parks. Relying on the construction of logistics parks and platforms, such smart platforms promote the orderly, intensive, standardized, large-scale, and modernized development of the industry.

3. The goal of the integrated construction of modern logistics and supply chain
To adapt to the development environment of modern logistics and fully utilize information technology, it is imperative to apply informatization at all stages of the supply chain. The logistics industry is developing in the direction of industrial technicalization, information industrialization, system intelligentization, and management integration. Based on electronic information technology, it is focused on the comprehensive integration of services, personnel, technology, information and intelligent management.

1.3.3 Demand analysis of the information platform

The smart logistics and supply chain information platform involves logistic, manufacturing, and commercial enterprise, public service agencies, and government departments. Its planning and construction require the support of IoT-related technologies and other information technologies built by the platform. Combined with the four aforementioned major platforms, it can achieve various functions as well full visual management throughout supply chain logistics. In addition, there are some non-functional requirements for its planning and construction. Therefore, the demand analysis of the platform is divided into four types of demand analysis – namely, user; functional; technical; and non-functional. The detailed analysis is presented in the diagram below.

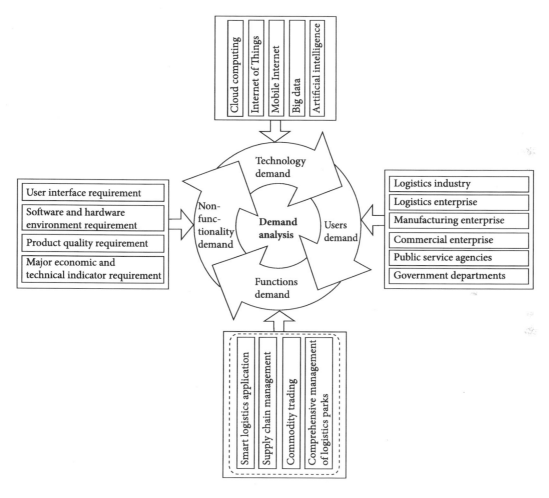

Demand analysis of the information platform

1. User demand analysis
1) Demand analysis of the logistics industry
The four core business areas of modern logistics are: integrated logistics; supply chain management; bulk commodity trading; and logistics park services. In order to meet a series of integration requirements such as information support, transmission, sharing, and resource integration,

targeted construction of the four major information platforms is necessary. Overall, information resources should be integrated, the concept of smart logistics should be injected into supply chain management and coordination, and then a comprehensive smart logistics and supply chain information platform should be created.

2) Demand analysis of logistics enterprise

Logistics enterprises are direct users of information platforms and key participants in their construction and development. They are mainly transportation, procurement, warehousing, and distribution companies that make use of the port logistics information platform. They require these platforms for collection and distribution information for freight, warehousing and distribution processing, bonded warehousing and customs declaration, as well as logistics value-added information. These enterprises have to obtain logistics supply chain services and customer resources through information platforms – to find other logistics service providers corresponding to the business. It is key for them to exchange data with other firms at every link of the supply chain, and to share information resources to bridge all logistics links.

3) Demand analysis of manufacturing enterprise

Manufacturing enterprises obtain information on short- and medium-distance distribution of goods from the information platform, such as receiving, warehousing, and outgoing goods. Information on the procurement, processing, transportation, warehousing, distribution, and sales of raw materials and on the purchase, sales and storage of various machinery and electrical appliances is also available on the platform. While obtaining this information, manufacturers can publish the above logistics service information on the platform, accept service requests, provide information about goods in transit, and exchange data with government departments and other companies.

4) Demand analysis of commercial enterprise

The commercial enterprise involved in the smart logistics and supply chain information platform is divided into service commerce and product commerce. They require platforms to provide information regarding warehousing, distribution and sales of products – specifically that of reliable distribution service providers, of local business transactions of the finished products, and of the large-scale product sales point, etc. They can, as a result, exchange and share data with other companies or institutions. Simultaneously, the smart information platform enables electronic transaction services, improves their operating efficiency and reduces costs.

5) Demand analysis of public service agencies

Banks, financial companies, insurance companies and other public service institutions need to strengthen business dealings through participation in platform construction, where they can render logistics value-added services such as online payment, utilize online insurance application and logistics finance, etc.

6) Demand analysis of government departments

(1) Government departments have demands for logistics management information too. They need to obtain information on the overall operation and distribution of logistics, the use of logistics infrastructure, and other types of information from the smart platform. Additionally, this platform can release government info on policies and regulations regarding state departments and competent organizations at national, industrial, provincial, and city levels to logistics companies and other related enterprise. They rely on the information platform to grasp and analyze logistics development at the macro level, to implement management and decision-making research to harness the information for scientific prediction, analysis and planning, and to formulate relevant policies to promote development.

(2) The government has demand for management information of the bulk commodity supply chain nodes. It obtains the basic, procurement, and sales information of the relevant enterprises listed in the bulk commodity supply chain through the information platform. It also manages information sharing on of the supply chain from a macro perspective to weaken the bullwhip effect.

(3) The government has demand for electronic bulk commodity transaction management information. It uses the information platform to extract transaction, market, and financial information during electronic bulk commodity transactions. As e-commerce information and products are displayed on the platform, electronic commodity transaction information is macro-managed, and relevant enterprises are urged to make micro-adjustments to ensure the circulation of fund flows between enterprise and reasonable profit distribution.

(4) The government has demand for logistics park management information. It releases public governmental information, shares public data and exchanges data between government departments, manages corporate credit, and supports industry decisions on the information platform. Simultaneously, it requires the platform to enable business subsystems of the logistics park to be interconnected, which strengthens coordination and cooperation, simplifies related review procedures, and enhances efficiency of functional management departments. The data collection and analysis by the information platform ultimately assists the government in decision-making.

2. Functional demand analysis

In the IoT application environment, based on the analysis of functional demand of the information platform as well as the functional characteristics of general logistics information platforms when combined with key business sectors of logistics, the functional demand of the information platform can be integrated with the four major business clusters of modern logistics. This integration realizes a variety of platform functions.

1) Smart logistics application function

To achieve integrated, informatized, and collaborative smart logistics applications, combined with the expansion of business content, the following requirements are made for platform functions:

(1) Application business requirements. In order to better serve platform-related production, trading, sales, and demand companies, the supply, production and sales logistics information of general, bulk, and dangerous goods is integrated in the transportation process, thereby improving the efficiency of integrated logistics services.

(2) Core business requirements. A variety of advanced technology is used to intelligently manage processes such as smart logistics in business control and supply chains, and global information control. The collaborative operation and management of core businesses are performed through the information platform, which improves the safety and quality of cargo transportation and the efficiency of warehouse operations. With the Internet and corporate intranet as network support, the center screen displays a plethora of information required by platform participants in the information service center halls and business exhibitions.

(3) Value-added business requirements. The IoT and middleware technology can realize intelligent processing, packaging, loading and unloading, settlement, automatic payment and accurate processing of real-time logistics status inquiries, thereby effectively reducing the management costs required for core business support while ensuring logistics efficiency and security.

(4) Auxiliary activity requirements. Big data technology is combined with corporate services to create services of collaborative supply chain management, logistics system design and optimization, and logistics process control. In addition, emergency management info, based on full-scale monitoring and visualized management of logistics information, is collected. This is so an emergency management plan can be formulated for business processes, and the safety of logistics firms is evaluated and timely managed. An early warning system was set up to reduce risks and improve safety in the business process. Emergency procedures and plans were also established to ensure the effective completion of logistical tasks.

2) Supply chain management function

For the purpose of collaborative management of the supply chain, business content expansion of logistics companies, and information sharing and unified decision-making requirements of users at each node, the following requirements are proposed for platform functions:

(1) Application business requirements. To better serve suppliers, manufacturers, sellers and logistical service providers at each node of the supply chain, a unified management system is built. Thus, the operation and information systems are closely coordinated to achieve seamless connection among all links.

(2) Core business requirements. Advanced technology is employed to render procurement, inventory, sales, and supply chain partnership management services for related companies at various nodes in the supply chain. In response to the diversity of demand for information platforms of various node companies, a multitude of services including business review management and business operation status management are used, improving a company's management efficiency of participating businesses.

(3) Value-added business requirements. To render value-added services including supply chain logistics, finance, and plan design to the nodes, the construction of an information platform, enabling the provision of logistical services to the supply chain node corporates, is required. And supply chain finance is conducted to alleviate capital pressure, reduce financing costs, and open up new sources of profit for the logistics industry.

(4) Auxiliary activity requirements. To better manage procurement, inventory, sales and supply chain cooperation, logistics companies require auxiliary support of credit management, information technology, performance evaluation, and settlement management. This in turn requires the platform to summarize and give feedback in a timely fashion on the above management processes. This forms the basis for unified decision-making of enterprise users at each node in the supply chain through the statistical summary of relevant logistics data.

3) Commodity trading function

To improve the electronic trading services of bulk commodities and simplify the transaction process, platform functions are required to achieve the following:

(1) Application business requirements. To facilitate the transaction of bulk commodities such as agricultural and sideline products, metals, and chemicals, the employment of an information platform is required. Such platforms can perform integrated information management and sharing for production and sales companies, corresponding demand companies, and related financial institutions.

(2) Core business requirements. The information platform trading system provides a virtual trading market for both suppliers and consumers. Offering personalized needs, it renders online logistics transaction services as well as commodity transaction services for both buyers and sellers. Digital communication networks and computers are adopted to replace the storage, transmission, statistics, and distribution of paper-based information carriers in conventional transaction processes, thus enabling online transactions, transaction management, settlement management, market information collection, and market analysis.

(3) Value-added business requirements. The platform integrates the information of a logistics company's loan business, advance payment business, investment and financing business, member transaction, and big data service. This makes its internal and external resources more effective. Consequently, high-efficiency logistics and fund flow, low-cost information management, and network operation are achieved.

(4) Auxiliary activity requirement. The information platform feeds users with logistics-related information. Users at different levels can publish pertinent information within the scope of their duties on the platform. They can search the logistics information needed via the Internet and wireless application protocol (WAP) from the platform. Thus, functions such as product information management and display, inventory management, and enterprise information management are realized.

4) Comprehensive management function of logistics parks

To meet the needs of comprehensive management informatization of logistics parks and realize seamless connection between local governments, logistics parks and enterprises, the following is required of platform functions:

(1) Application business requirements. The daily application business of a logistics park mainly includes information services for enterprise in the park and basic information services. The platform renders enterprise information on release, financial, talent, and material services. It offers information support to daily office work, enables office automation of the council of logistics management, and improves the management of logistics enterprises.

(2) Core business requirements. To optimize business and industrial services, the information platform, aimed at optimizing the needs of many links of logistics services (such as basic information, industrial services, e-government affairs, public services, and big data applications), should establish relevant mathematical optimization models. This can match services to departments through the optimization engine and improve logistic service quality. Meanwhile, it should be able to perform a statistical summary of relevant logistics data to support the macroscopic decision-making of government departments.

(3) Value-added business requirements. In response to a company's need for e-government services such as public government affairs, online announcements, and electronic approvals, technology, public commerce, and e-commerce, the platform should summarize, release, and display the information. This way, information sharing among various business systems on the platform and between the platform itself and external systems is encouraged.

(4) Auxiliary activity requirements. Big data and other information technologies are adopted to enable transportation, warehouse, manufacturing and commerce companies, as well as business management departments, to exchange information on logistics service requirements and capabilities. Via data analysis and mining, data combination, predictive analysis, smart business, operation analysis, and statistical analysis, a complete service solution is formed for public, industrial, and corporate needs.

3. Technical demand analysis

Networked, informatized, automated, and smart logistics – these are the characteristics on which the platform is based, including sharing and exchange needs of all platform participants. The technical demands of the platform can be divided into cloud computing, the Internet of Things, mobile Internet, big data and artificial intelligence.

1) Cloud computing

Cloud computing is a form of data-intensive supercomputing. It enables usable, convenient, and on-demand network access to a configurable computing which pools shared resources (resources are networks, servers, storage, application software, and services). These resources are readily available and require only minimal management work or interaction with service providers. The key technologies involved are virtualization, distributed mass data storage, mass data management, programming methods, and cloud computing platform management.

2) Internet of Things

The Internet of Things includes key technologies such as terminal, network, and information service technology. Terminal technology is adopted to perceive things, network technology to transmit and exchange things, and information service technology to render users with various types of information services.

3) Mobile Internet

Mobile Internet technology is the product of integration and development of Internet technology and mobile communication. It provides a mobile access network which contains three elements: mobile terminals, access networks and application services. The application network is at the core of this setup and is the ultimate goal for users. It is characterized by the ability to connect multiple heterogeneous groups that depend on each other to create combined value, to realize information docking between enterprise and users, and to bring a new model based on the platform and oriented to user demand. In this case, users are not only consumers of innovation outcomes, but also the source of innovation resource data.

4) Big data

Big data refers to an amount of data that is too enormous to be extracted, managed, processed, and organized by the current mainstream software tools within reasonable time. It has the characteristics of huge volume, high velocity, wide range, and great value. This technology can quickly obtain valuable information from various types of data. Its key technologies are generally the capture, preprocessing, storage management, analysis and mining, and visualization of big data.

5) Artificial intelligence

Artificial intelligence is a technology that employs machines to simulate human intelligence. It draws on the idea of bionics and abstracts knowledge in mathematical language to imitate biological systems and human intelligence mechanisms. It enables the logistics system to imitate human intelligence; thus it is able to think, perceive, learn, reason and judge, and solve certain problems in logistics all by itself. In this industry, the technical application of artificial intelligence mainly occurs in reasoning planning, mode recognition, and intelligent robots.

4. Non-functional demand analysis

Pivoting on the analysis of user, functional, and technical demand of the smart logistics and supply chain information platform, the planning and design of the platform are supplemented with non-functional demand analysis. This is presented as follows:

1) User interface requirement

The user interface of the information platform is the medium for interaction and information exchange between the system and users. It converses the internal code of information into an acceptable form for humans. A user interface is software that interacts and communicates between the users and hardware. It should enable them to operate the software with convenience and efficiency to achieve two-way interaction. An interface must always remain in touch, or relative, to users – heeding the feedback mechanism, and speaking in concise language with affinity. Operations should be easy, the interface should be beautiful, and all functions should be realized on the premise of guaranteed system safety.

2) Software and hardware environment requirement

The software and hardware environment is required to select applicable servers and for clients of the applicable system to reman compatible with the platform as well as visual programming and personalized programming designers to simplify user operations and meet individual customer needs. For example, Tersus (a development tool of management software customization) is an open visual application platform. Users can realize enterprise network application development by drawing a visual flow chart, instead of using a traditional programming code. Tersus allows users to develop all applications components, including graphical user interface, server-side scripting, business logic, and databases.

3) Product quality requirement

It is required that data input to and calculated by the platform should be free from deviation and there should be a reliable and accurate fault handling mechanism. When a record is deleted, added, or updated, the system response should be quick enough and easily deployed on relevant operating systems. Feedback and normal platform operation should be processed, corrected, and inspected in a punctual manner.

4) Main economic and technical indicator requirement

In platform design and development, the core indicators of platform operation should be considered. Based on big data, cloud computing and other information technologies, various platform functions are displayed through system integration technology. The primary technical indicators form the upper limit of user numbers, daily business processing time, comprehensive query time, reliability, scalability, cross-platform, and platform features, etc. Various technical indicators ought to meet the basic needs of users.

CHAPTER 2

BASIC THEORIES OF THE INFORMATION PLATFORM

The basic theories of the information platform serve as the theoretical and technical guidance for the construction of the smart logistics and supply chain information platform. They determine integration principles, plan integration tasks, and lay the theoretical foundation for information platform construction. The core of these basic theories lies in system integration, which includes integration elements and forms as well as integration design models.

2.1 An overview of information platform theories

In the information age, the establishment of a smart platform is an effective and important means to develop modern logistics. To maximize the value-added functions of this platform, it is necessary to effectively control and coordinate the flows of goods, information, and capital, bridge the borders of enterprise and industry, and to organically integrate the operation and industrial chains. With the aid of computers and modern management, multiple information systems with distinct functions have been integrated, becoming the theoretical core of the smart information platform.

2.1.1 An overview of System Integration Theory

System integration refers to the integration of people, technology, equipment, information, and processes related to computer hardware, software, and application objects. Processes and functions are integrated as hardware, software, technology, and information. Specifically, various personnel form a collaborative work team, adopting system engineering to rationally configure computer hardware, software, technology, information, and human resources in accordance with the special requirements of the application field. This optimizes a combination of management control and human-machine systems to enable automatic information processing, the formation of an application system that meets user requirements, and the achievement of high efficiency gains and general benefit.

2.1.2 System integration principles

1. Openness and standardization

An integrated information system requires openness. Only an open system can meet operability, portability, and scalability requirements. Standardization suggests that the selection of system hardware and software platform, communication interface, software development tools, and network structure should follow industrial open standards. Only open standardized information systems can be connected with other open systems; thus it can be continuously expanded and upgraded.

2. Practicality and advancement

The advancement of this system relies on the advancement of technology. Only with such technology can the advantages of the system and a long life cycle be ensured. Simultaneously, the system should be designed with special focus to practicality and close integration of the actual needs of specific applications. When certain network technologies are being selected, current and future mainstream applications should be taken into account; while blind pursuit of new technologies and products should be avoided.

3. Availability and economy

On the premise of meeting the system's functional requirements and achieving system construction, economic principles should apply, minizing total project investment and post operation and management costs upon completion. Availability means the system must have a fine and reliable performance to ensure that under any circumstances, the reasonable allocation of resources is achieved.

4. Reliability and safety

System integration should abide by these principles. In order to ensure data security and consistency, a variety of inspection and processing approaches must be made available. Corresponding security and reliability strategies are formulated for each level of the host, database, network, and application. It adopts fault-tolerant design, fault detection, and recovery technology to guarantee safety measures are effective and reliable, and safety control is feasible at multiple levels.

5. Flexibility and scalability

System integration has to be configured flexibly with alternatives and optional solutions ready. It should be able to expand its scale and performance to meet the requirements of technology and application development. The system must fully account for fine scalability and flexibility in structure, capacity, communication capabilities, product upgrades, processing capabilities, databases, and software development.

2.1.3 Tasks of system integration

The key to realizing system integration is to solve interconnection and inter-system operability problems. It is a multi-vendor, protocol and application focused architecture. All integration-oriented problems regarding types of equipment, interfaces, protocals, system platforms, application software, construction coordination, organization management, and staffing require urgent solutions. And the information system integration tasks can be described as six layers:

1. Integration of support systems

The integration of support systems is a significant foundation. In the support systems, enterprise application integration technology is mainly responsible for the completion of three types of task. The first is application integration. Via transmission and conversion, different application systems share information and data for mutual use. The second is internal process integration of enterprise. The heterogeneous and decentralized applications are effectively integrated as required by the operator's business process. The third is process integration of the business community park. Like the process integration within the operators, cross-organizational process integration extends integration to related companies and key customers across the entire supply chain. It effectively integrates applications and business processes between different companies.

2. Integration of information

Information integration aims to effectively integrate the information distributed in autonomous and heterogeneous local data sources, thereby enabling information sharing between each subsystem. Meanwhile, it must also resolve the problems of effective conversion among data, information and knowledge (including experience).

3. Integration of application functions

The demand for information determines the demand for integrated system functions. This integrates the functions of each application system as per a particular open protocol, standard, or specification under a unified framework. This become an integrated multifunctional system, where mutual calls and communication are realized; thus an integrated information system can better work its magic.

4. Integration of technology

Integration of technology is the center of the entire information system. Whether it is the realization of functional goals and requirements, or the integration between supporting systems, it is actually achieved by the integration of various technologies. This can be divided into hard and soft technology integration, and tool integration.

Hard technologies involve information like computer, communication network, database, data warehouse, and reusable software reuse technologies as well as management technologies such as simulation, prediction, and analysis technology.

Soft technology integration refers to the methods and models in information system integration, including system development and management method integration, such as object-oriented, structured, prototype, life cycle, and information engineering methods.

Tool integration refers to a collection of modules composed of multiple tools. It is often used to combine hard and soft technology into one unit to serve the management functions of an organization.

5. Integration of people

System integration must go through people, using a variety of hardware and software technologies to re-optimize and combine individual information systems, thus forming a unified and comprehensive system. The integration of people plays a key role in system integration. It should include the integration of people and technology, and human-machine collaboration. By its nature it is a human-oriented intelligent human-machine system. Therefore, it plays an important role in constructing integrated information systems and is the key to success of an integrated system.

6. Integration of products

The integration of products is the most direct manifestation of information system integration because whether it is an application function, supporting system, or technological type of integration, its ultimate manifestation is the integration of specific products. It is the external form of expression of system integration.

2.2 Elements and forms of system integration

The chief task of building a smart logistics and supply chain information platform is system integration research. Based on the distinct characteristics and focuses of its four sub-platforms (smart logistics information, bulk commodity supply chain service, bulk commodity electronic trading, and logistics park information service), its specific integration content is analyzed and studied. These elements of system integration are studied at both macro and micro levels, while its forms are dissected from the perspectives of both horizontal and vertical integrations.

2.2.1 Analysis of system integration elements

The smart logistics and supply chain information platform integrates logistics customers and service providers, logistics-related departments, and the socialized, open, and interconnected public information system of the supply chain. The integration of this platform does more than merely connect suppliers and logistics to users, but creates one integrated supply chain. Apparently, macro-level logistics, information, capital, and business flow, as well as customers, service providers and related departments, form the core components of the smart information platform, whose integration elements are displayed in Figure 2-1.

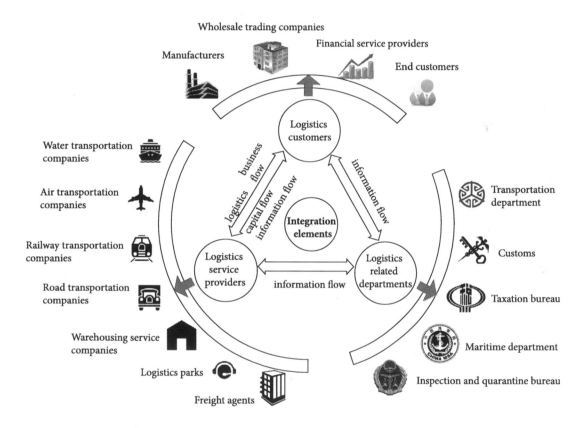

Figure 2-1   Integration elements of the smart logistics and supply chain information platform

1. Macro elements

The information platform is an online service website that operates with the Internet as a medium. It integrates business, information, and fund flow with logistics. The system technology enables the integration of industrial resources, logistics resources, and policy information, etc. The communication channel of these four elements is constructed between the subjects, paving the way for the construction of the smart platform.

1) Business flow

Business flow refers to the process of a transaction activity. The transfer of product ownership occurs during business flow activities. And this transfer requires emphatic reflection of business flow in electronic transactions, which notably includes neither payment nor transfer of funds.

2) Logistics

Logistics refers to the displacement of goods in space and time, including circulation links such as procurement and distribution, production and processing, and warehousing and packaging. As e-commerce develops, most goods and services adopt traditional logistical means, while a handful of goods and services can be directly distributed via network transmission. These include electronic publications and information consulting services. The logistics of the information age is characterized by informationization, automation, networking, intelligence, and flexibility.

3) Information flow

Information flow mainly includes the provision of product information, promotion, marketing, technical support, after-sales service, and commercial trade documents, as well as information regarding payment ability and transaction party credit, intermediary reputation, and a large amount of authoritative government publications. The smart platform management is based on the effective control of information flow within enterprise, between enterprise, and in relevant government departments. Enterprises employ modern information network technology to streamline information flow, while the government takes advantage of the interaction of to carry out safety supervision and decision-making support in the industry.

4) Fund flow

Fund flow concerns the flow of funds between marketing channel members as physical goods and their ownership transfer. In electronic transactions, whether the online payment made by customers can reach the payees safely, punctually, and conveniently determines if the transaction succeeds or fails. Therefore, online payments are of great significance for both customers and business owners. The key to this is the construction of a fund flow platform; of the four flows, it is the most important.

2. Micro-elements

The micro elements in the system integration are the carriers of macro elements. There are three basic logistic elements: customers, service providers, and related departments. The participants of this platform, driven by transaction demand through the segregation of duties and cooperation in the logistics process, achieve continuous value increase of the entire platform – the mediums are capital, logistics, business, and information flow.

1) Logistics customers

Logistics customers in the information platform refer to the companies or individuals with potential needs for services – including manufacturers, wholesale trading companies, end customers, and financial service providers – in the logistics process. A customer-centric management strategy is critical to this platform for attracting customers and nurturing their loyalty.

2) Logistic service providers

Logistic service providers are the core of the platform's integration elements. The upstream connects with product suppliers to ensure the safe and correct delivery of products; it also shoulders the responsibility of inventory management to reduce costs; the downstream contacts service end customers to ensures high-quality in the logistics links; meanwhile, meeting the reasonable information service needs and properly evaluating them is required. Finally, it must be linked to a bank to complete the payment and transfer of funds and provide logistics financial services. Water, air, railway, and road transport as well as warehousing service companies, logistics parks, and transportation agencies are examples of logistic service providers.

3) Logistics related departments

The smart logistics and supply chain information platform is an important means for the government to implement macro-logistics control. The industry covers a wide range of business; since it is difficult for conventional macro-control approaches to efficiently manage and support these, a smart public platform is introduced. Through accurate, real-time statistic gathering and analysis of the basic operating data, the logistics trading market is standardized and supervised – an information platform directly supporting enterprise development and policy guidance is built. And logistics-related laws and regulations are formulated to truly realize government functions under market conditions. For example, transportation departments, custom offices, tax offices, maritime departments, and inspection and quarantine bureaus rely on such platforms.

2.2.2 Analysis of system integration forms

In nature, the integration of this platform is a sort of resource integration between operations, enterprise, and industry, assisted by the value chain and the IoT. Driven by market competition and information technology innovation, enterprises pursue seamless links and organic collaboration of logistics as well as information and fund flow in the production process. This system integration is fundamentally embodied in the advent of inter-enterprise information integration, inter-industry collaborative network research, logistics service chain management, and value chain reconstruction.

These integration forms of the smart information platform are divided into horizontal and vertical:

1. Horizontal integration

Horizontal integration concerns daily operation systems between various departments within enterprise or between enterprise of similar level in the smart platform; it is the identification, selection, operation, and coordination of the same type of resources and the business system. This highlights the horizontal improvement of superior resources internally, which is epitomized in two levels.

Level one is the inside of enterprise which adopt a smart logistics and supply chain information platform. The most typical manifestation is the parallel engineering, on-time production, on-time procurement and logistics operation coordination among manufacturers, operating traders, and logistics service providers. They coordinate the business relationship between internal departments, combine similar resources, and establish close working contacts between departments.

Level two is between enterprise which adopt this smart platform. Typically, the franchised companies in the platform share advantageous resources in logistics on the basis of cooperation, forming a strategic alliance and building a community of interests to participate in market competition.

As per the core elements in its process, horizontal integration is divided into enterprise and organization integration. This is shown in Figure 2-2.

1) Enterprise integration

Horizontal integration mostly achieves collaboration between same-level enterprise and sold products. For manufacturers, the integration between upstream suppliers or downstream distributors of the same level or type fall within this category; for logistics service providers, those of the same business type and transportation service capabilities are qualified for horizontal enterprise integration; for companies themselves, the communication of their internal business information across departments is also a form of horizontal enterprise integration. This is divided into three models.

(1) The integration of macro-enterprise clusters centered on finance. Manufacturers or logistics service providers on the smart platform, whether loosely or tightly integrated in its sales end and supply chain system, support or draw plans for each other financially; in the context of large-scale enterprise or supply chain collaboration, this integration with the bank is feasible for macro integration.

Therefore, there are two main forms of macro enterprise integration: one revolves around "large banks + large enterprise"; the other is achieved by "multiple large enterprises" together.

The relationship between large groups or large logistics companies that are centered on smart logistics and supply chain information platforms formed by these two models is mutual competition rather than a monopoly. This kind of competition is mostly in the same industry and is moderate and benign. It is conducive to promoting the continuous improvement of products or service quality, price reductions, and consumer benefits.

(2) The integration of product-centered Meso-level enterprise clusters. This is the integration of enterprise clusters centered on products in the smart platform. A large group of manufacturers, wholesale trading companies, and professional logistics service providers are integrated through this platform. In this form, companies design the logistics service chain around

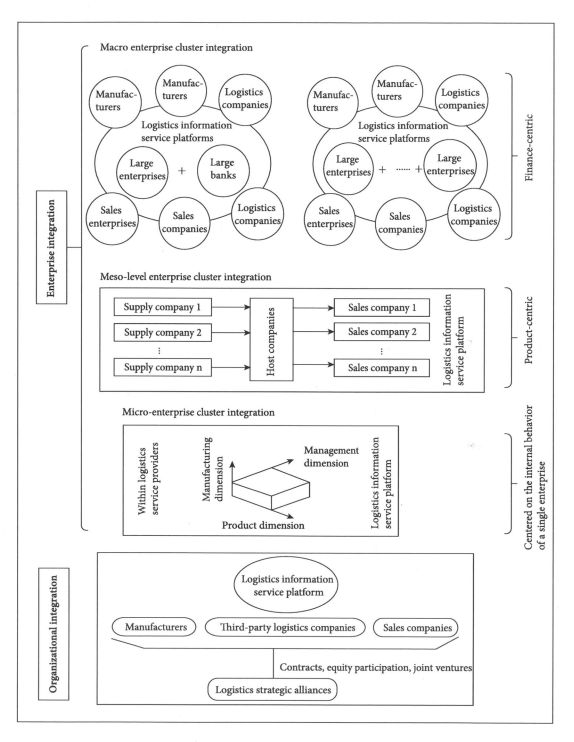

Figure 2-2 Horizontal integration of the logistics service information platform

a single product and extend it upstream and downstream. Centered on the integration of products, each company cooperates and supports one another in procurement, inventory, production, sales, and finance. As the information platform is closely integrated with logistics service providers, various operations of fast, sensitive, and smart logistics seamlessly connect every link.

(3) The integration of micro-enterprise clusters centered on the internal behavior of a single enterprise. This occurs as well as integration between product dimensions, manufacturing dimensions, and management dimensions.

2) Organizational integration

The organizational integration of smart platforms often refers to the co-establishment of the profit-sharing strategic partnership or logistics strategic alliance centered on online business. This features complete information, unobstructed channels, optimized configuration, specific roles, clear rules, risk sharing, and equity investment or joint ventures.

Organizational integration is more conducive to the formation of a multi-enterprise co-dependent and collaborative logistics service alliance with the smart platform at the core. This greatly enhances efficiency, optimizes resources allocation, acquirement, and utility, develops new resources, avoids internal competition, and results in collaborative operation. Consequently, the operating costs of the logistics industry fall while its performance improves.

2. Vertical integration

The vertical integration of the smart platform is the balance – revolving around the platform itself, market supply capabilities, production and manufacturing plans, and logistics service capabilities of franchised companies.

This integration style is able to reflect the service capabilities of the platform as well as the supply capabilities and willingness of upstream companies to the requirements of downstream companies. It strives to maintain coordination and synchronization. Judging from the depth and breadth of the smart platform, three layers can be observed in its vertical integration: information flow, business, and comprehensive logistics service chain. This type of integration is depicted in Figure 2-3.

1) Information flow integration

The logistics service chain is co-driven by business flow, logistics, and information and fund flow. For integration, it is necessary and plausible to start with these four resources – the most important of which is information flow.

This refers to the harnessing of modern information, database, multimedia, and system integration technologies to develop management software for the platform. It enables the collection, integration, analysis and processing of information in all aspects throughout the platform, makes forecasts, and assists with decision-making. Additionally, it supervises and controls intermediate service links in real-time, reduces overall management costs, and promotes management efficiency.

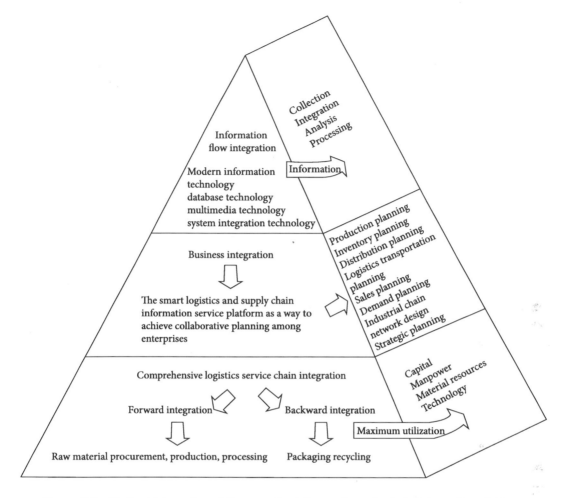

Figure 2-3   Vertical integration of the smart logistics and supply chain information platform

2) Business integration

The business integration of the smart platform is mostly epitomized in the implementation of collaborative plans between enterprise. Collaborative plans are the co-designs and implementation plans that member companies in the logistics service chain make in response to production plans, market demands, and work chain reactions as well as the production operation measures based on shared information.

These collaborative plans mostly involve production, inventory, distribution, logistics transportation, sales, demand, industrial chain network design and strategy.

They are implemented by synchronous product design, trial production, and large-scale customized production. The smart platform swiftly inputs personalized and customized orders and new product demand information into the system, thereby achieving uniform flexibility and high efficiency – a model of mass customization.

3) Comprehensive logistics service chain integration

The vertical integration of the logistics service chain is divided into forward and backward integration as per the extension direction. Forward integration means the enterprise holds full ownership and control over its integrated production. In such cases it is common for manufacturers to carry out all stages, from the procurement of basic raw materials to production and processing and assembly; or this is done by their own joint ventures or holdings, or outsourced companies; Backward integration, common in extended services such as packaging and recycling, is when companies control their customers.

On the whole, comprehensive logistics service chain integration is dedicated to combining forward integration and backward integration on the basis of partial integration as much as possible. It truly integrates the advantageous resources of the enterprise involved in the entire logistical process by giving full play to the advantages of information exchange, resource sharing, plan synchronization, and technology complementation.

As a result, the utilization of various resources such as capital, manpower, materials, and technology are maximized, while a well-integrated logistics service environment is created. The franchise companies on the platform are offered a brand new business operation model with which they can pursue business objectives in a brand new and more effective fashion.

2.3 The basis of system integration

The smart logistics and supply chain information platform effectively succeeds in the co-development of the manufacturing and logistics industries, connecting scattered customers, service providers, and logistics-related departments into a dynamic, integrated, and virtual network. The establishment of this smart platform is based on the collaboration of policies, industries, enterprise, technology, models and methods. The basic elements of information platform system integration are shown in Figure 2-4.

2.3.1 Policy basis

Recently, the government and relevant departments have formulated a series of logistics planning and development policies. These include the Medium and Long-term Planning for the Development of the Logistics Industry (2014–2020) and Guiding Opinions on Actively Promoting Supply Chain Innovation and Application. This improves the policy and and legal environment to a certain extent. Simultaneously, as the logistics industry develops rapdily, local governments have gradually paid more heed to logistics informatization and vigorously developed this smart platform.

1. China's logistics development planning

As the structure of the Chinese logistics industry is optimized, its transformation and upgrading has accelerated. Driven by both capital and technology, new business models keep springing up,

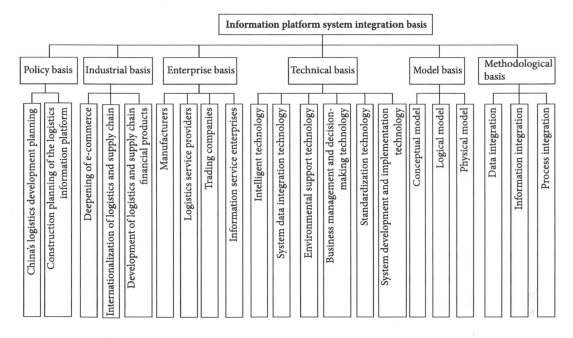

Figure 2-4 Basic elements of information platform system integration

platforms are integrated, and cross-border business taking place. This has profoundly affected and changed the traditional logistics industry.

In September 2014, the State Council issued the first plan (mentioned above), positioning it as a basic and strategic industry that pillars national economy development. It has reached new heights and a great space is given for development. In October 2017, the General Office of the State Council issued the second plan (mentioned above), comprehensively deploying work related to supply chain innovation and application, and propelling a general upgrade of the Chinese supply chain.

In line with the development trend of Internet +, the Chinese logistics industry has exhibited enormous vitality. With the widespread application of big data, cloud computing, IoT, mobile Internet, and artificial intelligence, new logistics models with Internet characteristics keep appearing. And Internet thinking speeds up the transformation of traditional industry as well as establishing a networked, intelligent, service-oriented, and collaborative ecosystem of Internet + efficient logistics.

2. Construction planning of the logistics information platform

In September 2014, the plan issued by the State Council set the smart logistics and supply chain information platform project as one of twelve key projects, requiring the integration of its current resources to form a cross-industry and cross-region public service platform.

In October 2017, the State Council's proposed the promotion of innovation and formation of a complete and efficient industrial supply chain. It is required to concentrate on quality and efficiency improvement of economic development, follow the path of deep integration of the supply chain and the Internet, and take support from informatization, standardization, credit system construction, and talent nurturing. Thus, innovating and developing new concepts, technologies, and models of the supply chain, efficiently integrating various resources and elements, enhancing industrial integration and collaboration, and buildinf a smart supply chain system underpinned by big data, networked sharing, and intelligent collaboration.

2.3.2 Industrial basis

Generally, in addition to the leadership of core manufacturers of the industrial chain, more is achieved by building an integrated supply chain service platform to render online and offline integrated services for related companies, such as centralized procurement and distribution, logistics services, platform transactions, and financing payments. This meets the needs of resource integration and function enhancement of various enterprises.

1. Deepening of e-commerce

As the application of modern information technologies such as the IoT, cloud computing, big data, mobile Internet, and artificial intelligence deepen in logistics, the physical operation of real-world and virtual (online) logistics information has begun to integrate in all aspects, pushing modern logistics into the era of smart logistics. The integration has propelled enterprise to apply lean supply chain management to perfect the entire supply chain system, so as to accrue informatization, services, and intelligence between the logistics and supply chain service network. It helps specialize and segregate labor and encourages collaboration within enterprise so that they can swiftly respond to customer needs, shorten production cycles and the time to launch new products market as well as reducing production, operation and transaction costs.

2. Internationalization of logistics and supply chain

With the proposal of the new Chinese regional economic strategies, adjustment of the Chinese logistics industry is progressing in an orderly manner. The Belt and Road Initiative emphasizes infrastructure interconnection, leadership of international logistics, and the implementation of logistics infrastructure construction and network layout. It aims to strengthen transportation hubs, logistics channels, information platforms, and other infrastructure construction. It catalyzes frequent contact with countries along the Belt and Road, establishes international cooperation in production capacity and equipment manufacturing, constructs economic cooperation zones along borders, cross-border economic cooperation zones, and overseas economic and trade cooperation zones. This encourages enterprise to deepen foreign investment cooperation, set up overseas distribution and service networks, and form a localized supply chain system.

3. Development of logistics and supply chain financial products

Sharing information is promoted among national and local credit information sharing platforms, commercial banks, and core supply chain companies. Commercial banks and core supply chain companies are encouraged to set up financial service platforms to render efficient and convenient financing services for small, medium, and micro enterprise upstream and downstream of the supply chain. Core supply chain companies and financial institutions are recommended to connect with the recievable accounts financing service platform built by the People's Bank of China Credit Information Center, in order to develop financial models.

2.3.3 Enterprise basis

Since the market demand is for real-time and personalized products, the integration of smart platform systems has become the trend of the times and a new business model for its efficiency, interaction, and adaptability. To cater to contemporary needs, manufacturers and logistics service providers have established this platform to broaden their own rapid development path.

1. Manufacturers

This smart platform is a virtual supply chain based on the Internet, gathering many logistic service providers, and breaking through conventional constraints on time and space. It is necessary to establish a standard entry system for manufacturers and a dynamic evaluation and feedback system. Also, punctual online communication, and online performance evaluation is carried out to harness compatibility between manufacturers and the platform itself.

2. Logistics service providers

Logistics service proviers are the most important enterprise basis for the smart platform. At present, Chinese logistics companies are mostly small and medium-sized. They are also still, numerous, scattered, and weak. Within them, there is commonly decentralized operation, a single function, little automation, unreasonable layout, limited technical content, weak horizontal cooperation with other enterprise, and poor service awareness and quality. The establishment of a smart logistics and supply chain information platform is one of the most critical approaches to improve industrial concentration. It is also a major opportunity for resource integration and structural reorganization.

3. Trading companies

Trade is the basis and condition for logistics production and development. Trading companies are more than just primary sources of trade or commercial activities; they are the main body of logistics activities. Currently, most Chinese commercial and trade companies are still at a less advanced development stage, which can include small-scale operation, single-service business, backward-circulation methods, unscientific management, and insufficient development stamina. They urgently require transformation and upgrade to adapt to the modern market economy.

Should they intend to meet market demand for small batches, multiple batches, multiple varieties, and urgency, they have to communicate with upstream manufacturers via the smart platform to maintain a stable goods supply. In return, market dynamics are fed back to the manufacturers in time for intensive production.

4. Information Service enterprise

Information services refer to the development and application of information technology rendered by the supplier to the consumer, and those from the supplier to support the business activities of the consumer by means of information technology. As a novel industry, information services cover a wide range. Meanwhile, with the innovation of information technology and business models, information service companies themselves are constantly developing and changing, resulting in a growing number of business and service types. Currently, the services they offer include logistics information on technical consulting and design and development, information system integration, and data processing and operation.

2.3.4 Technical basis

Integration technology is a fundamental support pillar for the smooth completion of the smart platform. Instead of simply interconnecting various technologies, it organically combines separated subsystems into an a single system with more powerful functions. This enables them to work in coordination, exerting overall benefits and achieving optimum performance. The system integration of the platform is based on intelligent, system data integration, environmental support, business management and decision-making, standardization, system development and implementation technologies.

1. Intelligent technology

Intelligent technology refers to the application of modern communication and information, computer network, industrial, and intelligent control technologies. This includes barcode, radio frequency identification, sensors, and global positioning systems (GPS). The intelligentization of logistics has become a direction of its development under e-commerce. It adopts advanced IoT technology to realize automated operation and optimized management of cargo transportation as well as improving the service level of logistics.

2. System data integration technology

System data integration technology includes data aggregation and data concentration technology. By applying them, system operations can be performed, and the establishment and maintenance of data access and integrity constraint rules is feasable.

1) Data aggregation technology

Data aggregation integrates multiple databases into a unified view. The data aggregation generated by the tool is a virtual database. Although it does not store any data, it contains the content of multiple physical databases.

2) Data concentration technology

Data concentration is used to perform schema mapping between databases through data conversion tools, and copy and convert data from one database into another, thereby gathering data from multiple databases into a single unified database. This has many benefits.

3. Environmental support technology

System integration requires the solving of all integration-oriented problems of various hardware equipment and application software that are related to subsystems, architectural environment, and staffing. These all require support of the environment. Such technologies include networks, databases, integration platforms/frameworks, computer-aided software engineering, and human/machine interface technology.

4. Business management and decision-making technology

The strategic planning of the entire system requires various decisions. Smooth decision-making is achieved when intentions are successfully conveyed and implemented. Business management and decision-making technology breaks the constraints of information services. It starts from the overall situation and provides decision makers with multi-angle and multi-level decision-making service support.

Enterprise modeling is a brand-new business management model offering a framework for companies to ensure their application system is perfectly matched with the business process that is constantly being improved; system development and implementation of technology lay the foundation for normal system operation. Data mining extracts from a large amount of incomplete, noisy, and random data are not commonly known yet they are potentially very useful for decision making.

5. Standardization technology

Standardized technology involves data exchange, process information standards, data format standards, and graphic software standards. Graphic software standards refer to the interface standards for data transfer and communication between system interfaces.

6. System development and implementation technology

Good system development and implementation technology can make development more efficient, reduce repetitive work, and keep developed products from deviating far from actual requirements. Computer-aided software engineering (CASE) provides an automatic approach for the design and documentation of traditional structural programming techniques. It is used for computer software development, operation, maintenance and management, and can assist software development at all stages while saving system development time. The data of this system is not completely public. For the sake of network security, data transmitted between the network interfaces of each subsystem must have certain security and confidentiality measures. During system development, reusing existing software elements as much as possible can speed up the process and improve software productivity on the basis of ensuring quality.

2.3.5 Model basis

The digital system model of the smart platform is based on the integration of physical, logical, and application systems. According to the platform and the major businesses involved in its upstream and downstream enterprises, its conceptual model is extracted and the application system is hierarchically decomposed. Each system and subsystem that it covers are desinged, and finally the entire information system is integrated through the physical and logical models.

1. Conceptual model

The conceptual model of the smart platform is a bridge between subjectivity and objectivity. It is a conceptual tool adopted to design a system and collect information for a certain goal. Regarding the computer system, the conceptual model is an intermediate level from the objective world to the machine world. First, the real world is abstracted into an information world, and then this information world structure is transformed into a machine world model. This then performs data and information modeling based on the user's views.

2. Logical model

The logical model is a description of the system's internal logical structure. In essence, it describes computer views of the conceptual model as a logical mapping of real-world information processing in the computer world. The logical model depicts all information details on the platform, including its interrelationship, operation, input, output and storage.

3. Physical model

The physical model is the physical realization of the smart platform's comprehensive integration. It describes the system's data processing, storage structure and network structure. The physical model is essentially the realization of the logical model in the database integration, which determines the physical storage and referencing strategy of data. It can improve data analysis efficiency and processing as well as shared database performance.

2.3.6 Methodological basis

For the comprehensive integration of smart logistics and supply chain information platform systems, in addition to the provision of structural framework references, a variety of methods are required as support. The platform system involves multi-function and multi-business integration, the application of comprehensive integration methods, and the integration of related business systems via data, information, and process integration.

1. Data integration

Data integration is the key and focal point for the integration of the whole digital system. As the smart platform business systems are based on distinct database management systems, the original basic data in each system differs greatly in semantics and structure, which makes it difficult for information sharing. Therefore, it is necessary to integrate data and information of the system for better cooperation, and to strengthen the resource sharing of different business information systems.

2. Information integration

As per the needs of digital information integration of the platform, the disordered information of various business systems should be organized into an orderly information set, which mainly includes four levels – the integration of data flow, information flow, information management, and information service – thus laying the foundation for process integration.

3. Process integration

Process integration involves using computer integration support software tools to efficiently and timely realize the sharing of data and resources. Collaborative work between various application systems are integrated with various isolated applications to form the smart platform digital system. Integration of business process and business logic (application integration) has ultimately led to process-oriented integration.

2.4 Design model of the information platform system integration

The core of the smart platform implementation is system integration. In turn, the essence of this is the adoption of computer software and hardware technology, and database, automatic control, and network communication technologies, which optimizes the general system design for best overall performance. The construction of this platform should be based on the system integration model structure, while the system integration method is used to gradually build the conceptual, logical and physical model.

2.4.1 Construction approach of the information platform integration model

1. Modeling structure

The prerequisite for constructing a system integration model of the smart platform is the presence of comprehensive analysis and design framework. This framework is the integration modeling structure which describes the basic configuration and connection of each component on the platform. Basically, it opts to divide and conquer the complex system, analyze and describe the various sub-links of the logistics service chain from different aspects involved in company business, and treat the entire system as something composed of logical, physical, and application systems. Consequently, it forms a subsystem integration architecture. This is shown in Figure 2-5.

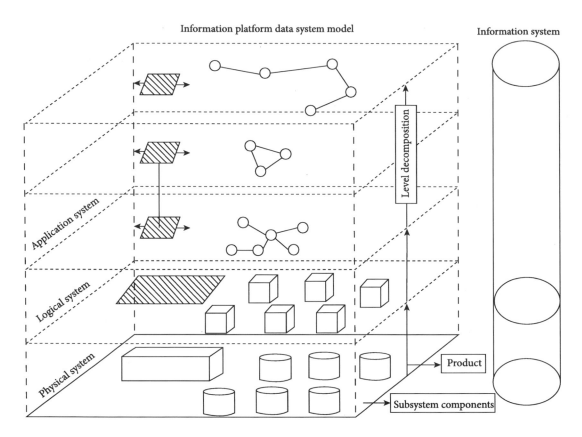

Information platform data system model

Information system

Figure 2-5 Information platform data system integration modeling architecture

This structure is based on physical, logical, and application systems. According to the principal business of manufacturers and logistics service providers on the platform, the data system is abstracted; next the application system is hierarchically decomposed while the data is analyzed and the systems and subsystems it covers are designed. Lastly, the integration of the entire information system is done through physical and logical systems.

2. Research on the approach system of information platform integration

This modeling structure of the information platform requires both a comprehensive integration modeling system for a structural framework reference, and a variety of data-based system comprehensive integration approach systems for support. Applying the latter it is possible to integrate related business systems through data, information, and process integration. This approach system of the smart platform data system is demonstrated in Figure 2-6.

1) Data integration

Data integration is quintessential for data system integration. Since the business systems of the smart platform are based on different database management systems, the original basic data in

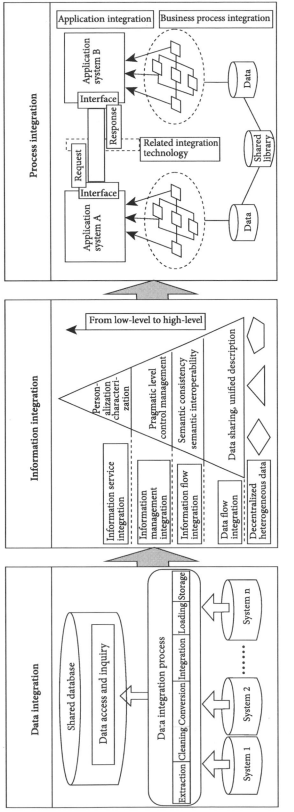

Figure 2-6 Comprehensive integration approach system of the smart logistics and supply chain information platform data system

each system differs greatly, making it difficult to share. It is necessary to integrate and pool system data and information while strengthening the resource sharing of different business information and heterogeneous systems.

For heterogeneous database systems, two things must be done to realize data integration: database conversion via data extraction, update, and storage; and transparent access to data, which enables business system data synchronization and sharing. For these requirements, data integration related technology is adopted to establish a shared database.

(1) Data extraction. Data extraction refers to the data processing that occurs before entry to the shared database. It involves the comprehensive integration of a shared database with unified storage from multiple data sources. It is responsible for extracting the data that needs storing from different business systems, performing data conversion, reorganization, and standardization of data from multiple sources (which vary vastly), and storing it in the database in a unified format.

(2) Data update and storage. A data update can involve cleaning, converting, integrating, loading, and refreshing data from the business database of the platform into the shared database. By setting the update frequency, the shared database can complete synchronous updates at custom regularity. Therefore, it can adopt data warehousing technology to form multi-level granular storage data while also forming multiple divisions according to the correlation of data access. Consequently, the storage of real-time data and historical data is complete, ensuring accuracy and avoiding data redundancy.

(3) Data access and query. The system adopts a database access technology called Open Database Connectivity (ODBC), which offers a common interface for accessing different databases. An application can access these different database management systems through a common set of codes. The shared database renders information services to the system according to service requests and query authorities, directly organizing and outputting the data it has stored. It can engage a combination of Web browsers and clients to render data query services according to user classification and needs.

2) Information integration

As required by the data system information integration of the smart platform, disordered information of various business systems is sorted, organized, and merged into an orderly information set. This includes the integration of data flow, information flow, information management, and information service, thus paving the road for process integration.

(1) Data flow integration. Between different business systems, the smooth flow of data is achieved through network communication protocols. The data is rationally organized, its exchange format is unified, redundant data is merged, and data free sharing is plausible between business systems.

(2) Information flow integration. Closer to intelligent interaction, it allows the interaction of different businesses to avoid the interpretation of complex data organization logic at the data level. From the application perspective, it shields the physical storage management of the underlying data.

(3) Information management integration. By creating a complete information management system, the effective operation of the information system is ensured along with the effective management of shared information flow. Information management is both static and dynamic, but dynamic management is of greater importance.

(4) Information service integration. Through various channels, information resources are collected, sorted and classified. And according to the individual needs of customers, different information services are rendered and recommended to users. The users receive personalized information services with information interaction between distinct businesses and users.

3) Process integration

Process integration is the adoption of computer integration support software tools to efficiently and punctually share data resources and collaborative work between various application systems. It also integrates isolated applications to form a coordinated data system. The integration of business processes and business logic (application integration) leads to process-oriented integration.

(1) Business process integration. This is integration of the data system through a process-oriented system in which the application system is managed. It includes business management, business simulation, and a workflow that combines any process, organization, and incoming and outgoing information. Since it spans multiple business process layers, message intermediaries are usually employed for integration. They use a general or hub model to standardize message processing and control the flow of information.

(2) Application integration. This revolves around the sharing of system functions, that is, business logic. The data system of the smart platform achieves application-to-application integration in a network environment via application programming interfaces. The goal of application interface integration is the reuse or mutual reference of functions between distinct applications.

Combining the integration requirements of the data system, the above modeling structures shall be adopted as modeling ideas, employing a comprehensive integration approach system to guide the construction of a highly integrated conceptual, physical, and logical model of the data system that is suitable.

These three models gradually abstract the smart platform system in reality. First, they describe the structure, activities and functions from a user perspective. Then, based on the process and data, they express the logical relationship of the integrated system through the connection between the processes. Eventually, they construct all the implementation details of the processing, storage, and network structures of the integrated system. This in turn guides the construction and operation of the data system.

2.4.2 Research on the conceptual model of information platform integration

The conceptual model is a bridge between subjectivity and objectivity. It is used to design systems and collect information for certain goals. Regarding computer systems, it is an intermediate level between the objective world and the machine world. It is, in fact, the modeling of data and information according to the user's point of view.

1. Analyze user subjects and needs and determine system service boundary

The primary users of the data system of the smart platform are the main receivers of the service as well as the subjects designated by a certain range of service – a fact that is the premise and basis for demand analysis and definition of user services. The user subjects are mostly manufacturers, wholesale trading companies, financial institutions, logistics service providers, and end customers.

A demand analysis of the user subjects is performed to explain what system functions and characteristics the users require from their perspective. The attainment of user needs lays the foundation for defining user services and sub-services. Additionally, the bridge that connects user needs and system functions is the conceptual model, which is also the basis for establishing the logical and physical models.

2. Define user services and sub-services and divide service subject domain

Considering the core business of the smart platform and actual user needs, the data system of this platform is divided as follows: industrial resource integration; logistics business services; online transaction services; collaborative operation services; administrative supervision services; and decision support services. The range of services here is shown in the following table. In addition, the system has to realize integration within and between the various service systems to achieve intelligentization and datamation.

The Range of Services of the Information Platform Data System

Number	Range of service	Sub-service
1	Industrial resource integration	Commodity supply and demand information, vehicle supply and demand information, logistics center and route information, etc.
2	Logistics business services	Inbound and outbound management, transportation management, warehouse management, dynamic tracking, etc.
3	Online transaction services	Transaction management, payment management, service evaluation system, etc.
4	Collaborative operation services	Collaborative operation management, business agency, etc.
5	Administrative supervision services	Policies and regulations, announcements, qualification reviews, industry supervision, etc.
6	Decision support services	Daily inquiry, industry analysis, statistical analysis, macro-control, etc.

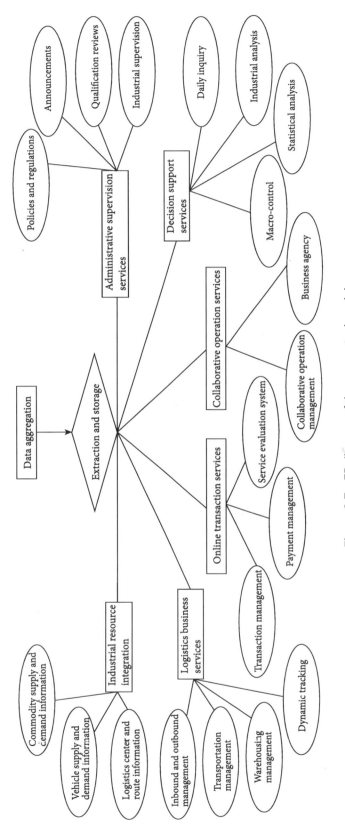

Figure 2-7 ER diagram of the conceptual model

3. Design the conceptual model of the system according to the service subject domains
Based on the service subject domains above, a rough system boundary for the moment can be delineated. In a sense, the work to define the system boundary can be regarded as the demand analysis of the design of various business and database systems. The requirements are expressed in the form of the system boundary definition, which is reflected in the functional requirements of the database system and the six service systems. An ER diagram of the conceptual model is shown in Figure 2-7.

The conceptual model of data system integration of the smart information platform completes the analysis and demand of the main body while dividing 6 service subject domains (see table above). It reflects the functional requirements of the system through data collection and sub-services of each subject domain. Also, it closely links user requirements with system functions.

2.4.3 Research on the logical model of information platform integration

The logical model of the data system describes the internal logical structure of the system. By nature, it is the description of the computer perspective of the system's conceptual model and the logical mapping of real-world information processing in the computer world. This logical model describes all details of the information on the platform and the relationship and operation between this information as well as processing information input, output and storage. In system integration modeling, the logical model is usually based on the process of data.

1. Establishment method of the logical model
The logical model determines the system function as required by the conceptual model, and reflects the analysis of field activity dynamics and job completion in the form of data. The establishment of the model tremendously relies on the Data Flow Diagram (DFD), a structured process modeling tool for description and decomposition which can demonstrate data flows and changes with direct diagrams.

The entity is the source or end of the data flow, which is expressed by boxes. It defines the boundary of the system, sending information to the subsystems or receiving it from them. DFD simulates the processing procedures of the problem from outside to in, and from top to bottom. After a series of decomposition steps, it elaborates the internal relationship of the system.

2. The design of the logical model
The logical model of the comprehensive integration data system aims to describe the relationship between system functions in order that entire system functions and data flow between functions are well organized. The design is divided into 3 steps:

1) Establish a functional hierarchy table as per functional demand analysis
Functional demand analysis is, from the perspective of the system, about organizing and summarizing the functions required to render services, merge similar functions for different

user subjects, and divide service domains into functional domains. This is fundamental for the construction of a comprehensive integrated logical framework.

2) Establish a data flow diagram of the system based on the functional hierarchy table

Data flow entity and data flow between systems are determined according to the relationship between the system and the external environment.

3) Adjust the data flow diagram

As per the function hierarchy table, each processing function and data flow goes through refinement so that connections between the decomposed sub-functions are relatively loose and

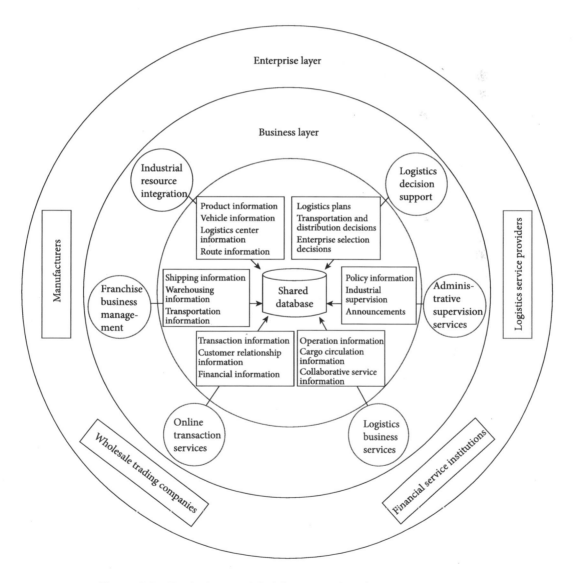

Figure 2-8 The logical model of the comprehensive integration of the information platform data system

simple. However, those between various parts of the sub-functions are tight and complex, and the balance of the data flow diagram is well maintained. The logical model of the comprehensive integration is shown in Figure 2-8.

This logical model of the data system designs each application under the six systems mentioned above. It finishes the integrated design through data exchange and integration between shared databases and various application systems. Figure 2-8 reflects the system's processing of the data flow. These processes are expressed as the functions performed by each system. Each subject in the box is the source or end of the data flow, which defines the system's boundary. They send information to each subsystem or receive it. Next, each subsystem stores the data in the shared database according to 6 different themes, thus fulfilling integration functions of each subsystem.

2.4.4 Research on the physical model of information platform integration

The physical model is the physical realization of the comprehensive integration of the data system platform. It depicts data processing and storage structure and network structure of the system, and has all its implementation details. The physical model based on data, information and process integration is essentially the realization of the integration of the logical model in the database, which determines the physical storage and reference strategy of the data. It improves the efficiency of data analysis and processing as well as the performance of the shared database. And it materializes the functions and data flow in the logical model into each system and subsystem, thus comprehensively designing entire system integration from the three aspects of infrastructure,

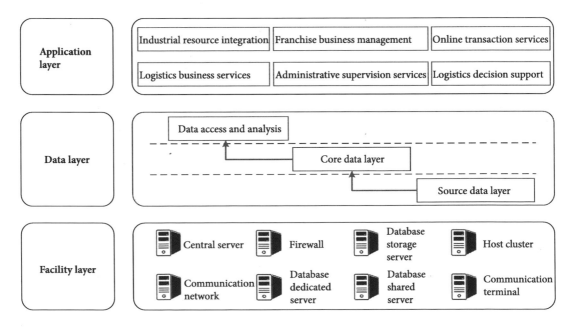

Figure 2-9   The physical model of the comprehensive integration of the information platform data system

data flow and application system. The physical model of the comprehensive integration is shown in Figure 2-9.

1. Facility layer analysis

The facility layer includes servers, storage, networks, firewalls, and host clusters that support the entire system. It should be designed with the capabilities of data transmission and network service between systems, the enhanced collaborative work between various business systems of e-commerce and higher communication efficiency between enterprises, and stronger need for the security maintenance of basic data. The system infrastructure of the smart logistics and supply chain information platform consists of system center server, system communication network and firewall, dedicated server and corresponding storage of the database of each business subsystem, shared database server and related storage, the host cluster of each application system and the communication terminals of various workstations, monitoring stations and monitoring equipment. These infrastructures are integrated and constructed in accordance with the needs of the system and the connection between business and functions.

2. Data layer analysis

Data sharing and integration technology are adopted to establish a shared database for the sharing of heterogeneous data, including the source data layer, the core data layer, and the data access and analysis layer. In the smart logistics and supply chain information platform system, the source data layer is the internal database of each application system. It provides basic and business data of the various business systems of the information platform while obtaining the data required from the shared database. The shared database is the center of the data layer. On the one hand, shared data is extracted from the source data and updated. On the other hand, it provides source data for the theme database and application system. The data access and analysis layer presents the data of each theme database to the application system in multiple dimensions by data mining and access, thereby integrating each application system.

3. Application layer analysis

The application layer integrates industrial resources, franchise enterprise management, online transaction services, logistics business services, and administrative supervision services and logistics decision support. The facility layer is the infrastructure for this action while the data layer supports it with data and information. The business of the application system is directly related to the subject area of the shared database, whose theme analysis is oriented at the application layer. Each theme library determines the business of each subsystem. On top of data sharing, each application system is interfaced as needed for their integration.

The physical model of the comprehensive integration of the data system of the smart logistics platform guides construction and application. It provides the integration approach for heterogeneous data of each system, thus achieving high integration and data sharing, and finally the collaborative work of each system and the application to which it belongs.

CHAPTER 3

BUSINESS CLUSTER AND SYSTEM DESIGN OF MODERN LOGISTICS AND THE SUPPLY CHAIN

3.1 Business cluster planning of modern logistics and the supply chain

The modern logistics industry is booming globally. For many nations, it has become the artery and focal industry of national economic development. During the development of modern Chinese logistics, various techniques and related theories are under constant exploration. These consist of transportation specifics, distribution, circulation processing, and information delivery. A large volume of new information keeps emerging in the market, which brings about rich opportunity as well as risk. However, the failure of companies in the supply chain to capitalize on this information in a timely and accurate manner hinders them from making correct decisions.

To achieve logistics and supply chain integration, these companies are supposed to fully share information and eliminate supply chain uncertainty. To do that, they must implement informatization and platform sharing, and design four major business clusters based on present industry development. The four major business clusters are integrated logistics, bulk commodity supply chain service, bulk commodity trade, and logistics park service. The division of these is shown in Figure 3-1.

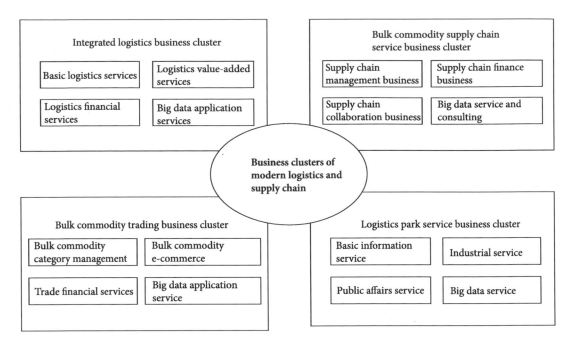

Figure 3-1 The division of business clusters of the modern logistics and supply chain

3.1.1 Integrated logistics business clusters

Integrated logistics refers to the process – in order to meet customer demand – of planning, implementing and regulating the efficient flow and storage of raw materials, products in progress, finished products, and related information from the source to the point of consumption. Chinese integrated logistics performs integrated management of the entire supply chain, comprehensively managing basic services such as transportation, warehousing, and distribution through process informatization.

The development of integrated logistics business clusters is closely related to that of basic logistics services. On the basis of continuously consolidating this business cluster, existing physical logistic advantages are relied upon such as road, rail and water transportation. Also, container multimodal transportation to consolidate and improve basic logistics services is needed. Modern management techniques and concepts are adopted to reengineer business processes, ultimately achieving value-added services centered on transportation, warehousing, information and the supply chain. Logistics service models are innovated to cultivate business centered on financing and warehouse receipt pledges, and to improve the corporate capital chain; while keeping up with the times, big data application business is carried out between the Internet and the company's intranet for intelligent management and control.

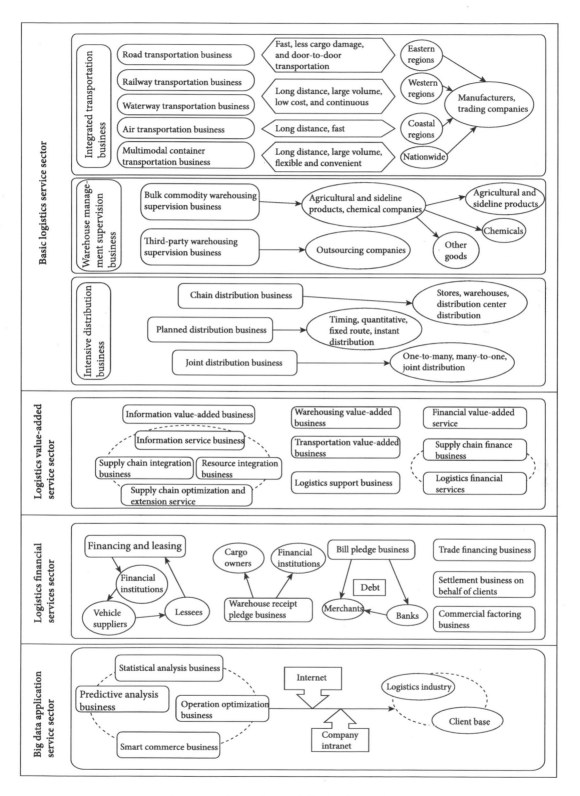

Figure 3-2 Integrated logistics business cluster

The integrated logistics business cluster carries out physical business with basic and value-added services as the focal point. The virtual logistics business revolves around logistics finance and big data application services, thus forming a dual-track operation mode with the physical. This contributes to the effective connection and operation of the four sectors and promotes smart logistics. This is detailed in Figure 3-2.

1. Basic logistics service sector

Basic logistics services refer to conventional logistics services. The central business of modern logistics is transportation, warehousing, and distribution, which is also a big part of basic logistics and hence the integrated logistics business cluster. This has advanced management levels, wide industrial chain coverage and strong comprehensive supporting capabilities.

1) Integrated transportation business

The integrated transportation business is also one of the basic logistics services as well as a significant link in the realization of logistics. It consists of the road railway, waterway, air, and multimodal container transportation businesses. These five primary transportation styles form an efficient system through collaboration.

(1) Road transportation business. This refers to the use of roads as the transportation line and the use of land transportation vehicles such as trucks to transport goods across regions or countries. As one of the primary transportation styles of foreign trade and domestic cargo flow, it is not only an independent transportation system, but also a crucial means for the collection and distribution of materials at stations, ports and airports. In remote and economically inferior areas with rugged terrain, sparse population, and underdeveloped railway and water transportation, roads provide the main (and often only) transportation artery.

(2) Railway transportation business. Goods are transported by train: it is a method generally not susceptible to climatic conditions and guaranteed to maintain normal operations throughout the year, employing railway facilities and equipment to transport passengers and goods. Aside from a high degree of continuity, it is also positively characterised by large capacity, fast speeds, lower freight, accurate transportation, and less risk. For medium and long-distance bulk cargo transportation and large-volume passenger transportation, it is often the optimum form of transport.

The subject of railway freightage is to properly arrange its equipment and realize loading and unloading mechanization; predict volume, and prepare and implement monthly freightage plans; scientifically organize the source and flow of goods, and extensively carry out direct, reasonable, and balanced transportation; arrange daily work; formulate conditions; load transported goods; manage railway freight yards and special lines; analyze freight accidents, and propose preventive measures.

(3) Waterway transportation business. In this method ships as the key transportation means, ports or port stations are the bases, and water (including oceans and rivers) is the transportation 'line'.

In many countries, waterways are of paramount importance as key transportation networks.

As the earliest and longest-running method, it has the technical and economic characteristics of large load capacity, low cost, and small investment; however, it performs poorly in flexibility and continuity, thus is more suitable for medium and long-distance transportation of large quantity and low value – heavy bulk cargo. Sea transportation, in particular, is ideal for this.

(4) Air transportation business. Air transportation uses airplanes as a means for cargo transport via air routes and airports. In China, it occupies a small portion of national freight volume transportation. It is well-suited to long-distance, short, light, and time-sensitive cargo transportation, and long-distance passenger transportation. It is characterized by great speed and strong serviceability.

(5) Multimodal container transportation business. This method adopts at least two different modes of transportation by the multimodal transport operator, transporting goods from the take-over location in one country to the designated delivery destination in another. It is a kind of continuous and comprehensive integrated cargo transportation that takes containers as the transportation unit, and organically combines distinct transportation methods.

Following the 5 requirements of: one carrier; one commission; one quotation; one charge; and one order, carriers in each transportation section jointly complete the entire transportation process. To some extent, modern logistics is developed on the basis of container multimodal transportation. Due to this, it forms an organic body with storage, loading and unloading, handling, packaging, circulation processing, distribution, and cargo information tracking links. It has increasingly made use of its unique advantages in modern logistics, thus playing a greater role.

2) Warehouse management supervision business

This refers to the supervision and management of warehouses and the materials stored in them. In modern logistics, warehouses are not only deemed part of the added value process, but also as a key factor in the successful operation of enterprise. Faced with new economic competition, companies (while focused on developing their own competitiveness) have paid greater attention to proper warehouse management. Precise warehouse supervision can effectively control and reduce circulation and inventory costs. The role of modern warehousing is more than just storage, but is also a material circulation center; its focus is now directed more towards technological utilization, such as information automation technology, to improve speed and efficiency. It is why automated warehouses are widely used.

(1) Bulk commodity warehousing supervision business. This business concentrates on agricultural and sideline production, and carries out corresponding warehousing supervision of bulk commodities. It integrates the management of bulk commodity inventorys, inbound and outbound warehouses, and customer statistics, etc. Through the rapid response and feedback of information absorption, data processing, and instruction transmission in various operating procedures of warehousing, it implements control, statistics, and effective data analysis of bulk

commodities. As a result, warehouse operation efficiency is improved, product loss is reduced, and losses arising from untimely replenishment are avoided.

(2) Third-party warehousing supervision business. This involves companies outsourcing warehousing to specialists, in which external companies provide comprehensive logistics services while the companies provide customers with storage supervision services. It can offer customers a complete set of logistics services, including storage, loading and unloading, and order processing. Throughout the entire process, the warehousing business is combined with IoT-related technologies, and the operation management platform is utilized to realize automatic warehousing, smart inventory control, etc. Outcomes include fewer unnecessary short deliveries, lower costs, better work efficiency, and higher operative efficiency.

3) Intensive distribution business

This refers to the selection, processing, packing, assembly, and distribution by various logistics companies, according to customer needs within economically reasonable areas. It also concerns intensive logistics services that deliver goods on schedule to designated destinations with optimized quality. Intensive distribution is supposed to be integrated across industries, departments, and regions on a national scale, thus forming a networked, informatized, social, automated, integrated, and multi-functional distribution system. It comprises of three business parts: chain distribution; planned distribution, and joint distribution.

(1) Chain distribution business. Chain distribution business targets stores, warehouses and distribution centers. As required by customers, they stock up and deliver. This is an economic activity organized and arranged according to user requirements. A complete chain distribution business not only enhances services, but is also a novel logistics method system reliant on modern information technology.

(2) Planned delivery business. Planned delivery is aimed at customers in the urban areas of different regions, providing customized delivery services. The customized services are fixed-time, fixed-quantity, and fixed-route deliveries. Oriented at various customers in the city, it coordinates transportation resources to avoid waste and render distribution services for the items in need.

(3) Joint distribution business. Also known as shared third-party logistics service, it means that multiple customers are united and jointly provided for by a third-party logistics company. This is carried out under a unified plan and schedule of the distribution center. There are many forms of joint distribution in modern logistics, namely, one-to-many, many-to-one, and co-delivery. Joint distribution improves logistical operation efficiency and reduces operating costs; it enables companies to concentrate on running their core business and promoting growth; market scope is therefore expanded, the old closed sales network is eliminated, and an environment of co-existence and common prosperity is built.

2. Logistics value-added service sector

Logistics value-added services have multiple merits including better convenience, faster response, lower total costs, and extended corporate services. At present, China's logistics market is going through a marked shift, where basic logistics profitability is gradually shrinking and the supply of value-added logistics services is terribly insufficient.

The logistics industry ought to grasp the opportunity provided by a relatively complete logistical foundation and brand effects in various regions, and strive to provide customers with personalized value-added services. The value-added service sector, a powerful supplement to the foundation of integrated logistics, can offer clients with other services. It involves all processes from product procurement to transportation, warehousing and storage, outer packaging, distribution, and information processing. The value-added service sector of the logistics industry consists of multiple businesses, such as transportation, warehousing, information, and logistics, which cover all aspects of production and life as well as contemporary financial value-added services.

1) Transportation value-added business

This relies on the integrated transportation business of the logistics industry to render customers with a number of value-added services. These include transportation route development design service, freight plan design service, advanced freight services, vehicle refueling and maintenance services, vehicle rental, and road rescue service. This ensured the efficient realization of transportation and distribution services. It aims to improve service convenience and rapid response while reducing costs.

2) Warehousing value-added business

This refers to the fact that the logistics industry depends on the basic business of warehousing supervision to render a variety of value-added services such as optimized packaging and processing. The logistics industry can cooperate with customer marketing plans to optimize the packing or product portfolio services of the stored goods, so as to improve the quality. In addition, it can also provide customers with quality inspection services, simple processing services, and product tracking and inventory inquiries to meet their diverse needs.

3) Information value-added business

Here the logistics industry relies on advanced information technology and open service concepts to integrate information technology into logistics operation arrangements should the conditions permit. It utilizes accumulated effective data to provide customers with information value-added services, such as, data query and customized services via network technology. Its operation can both predict customer needs more accurately and perform effective resource allocation to realize the total electronic management of goods.

4) Logistics support business

The logistics support business is the high-quality and thoughtful value-added service made available to meet all needs of production and life, such as the need for permanent and temporary offices, meeting rooms, catering and takeaways. There are cultural and sports value-added services too, which cover a range of of activities, as well as various support services such as accommodation. It not only ensures the normal operation of enterprise, but also serves their customers and generates new profits.

By launching value-added service business sectors, logistics companies can provide customers with a series of high-level, serialized, and fully-processed services. This ensures the efficient realization of basic logistics services (as mentioned above). Meanwhile, the logistics service model is further developed, enterprise profit margin expands, and competitiveness is enhanced.

5) Financial value-added services

Banks are often required to participate in the settlement and financing services rendered by logistics suppliers to customers. This new type of financial service is a type of derivative service. It is called a financial value-added service instead of a mortgage or pledge because as it develops, the authority-responsibility relationship between the banks and the loan applicant is completely different from the third-party relationship in which the guarantor assumes joint liability in a secured loan. Financial value-added services include supply chain and logistical financial business.

3. Logistics financial service sector

This is the operation process of logistics which effectively organizes and allocates the movement of monetary funds by applying and developing various financial products. These movements come in the form of deposits, loans, investments, trusts, leases, mortgages, discounts, insurance, securities issuance and trading. All types of intermediary businesses involving logistics are handled by such institutions.

Nowadays, the logistics industry has set up a variety financial services, such as, trade financing, leasing, mortgage guarantees, and warehouse receipt pledges to occupy a good position in the economic zone. This consists of financial leasing, warehouse receipt pledges, trade financing, customer settlements, and commercial factoring business. Obviously, perfecting the sector has become the key to the monetary activities of enhancing the capital chain, improving the capital turnover rate, and effectively organizing and allocating monetary funds in logistics.

1) Financing and leasing business

The logistics industry relies on basic logistics services to engage in the financing and leasing business of high-end equipment and vehicles for large companies, state-owned enterprise, truck owners and other customers. Hence, the financing and leasing business. It aims to relieve their fund pressures and meet their (lessees) personalized needs for designated manufacturers, equipment, and prices. It strives to vigorously develop equipment leasing and adjustment businesses. It can also receive support from the big data application service sector to engage in beneficial activities.

2) Warehouse receipt pledge

The warehouse receipt pledge business concerns financial activities performed during the temporary storage of goods in the warehouse. This gains new profits when the logistics industry depends on warehousing supervision and other business. As a basic financial business, it forms the internal demand for the development and innovation of enterprise seeking new profit growth points, and is a key component of the warehousing industry. It creates space for logistics to expand its service scope and perform diversified operations.

3) Trade financing business

These are the specialized trade finance activities of bulk commodity when the logistics industry owns a direct stake in small and medium-sized commercial banks through share expansion and capital increase. In commodity transactions, commercial banks can use structured short-term financing tools based on the financing bulk commodities such as inventories, prepayments, and receivable accounts.

4) Settlement business on behalf of clients

The servicing settlement business is the unified settlement management of the trading center, where any transaction made must go through the settlement of its platform. It has several modes such as delivery collection and advance payment.

5) Commercial factoring business

Commercial factoring renders financing services to small and medium-sized enterprise in the logistics industry, specializing in trade financing tools for comprehensive financial services such as receivable accounts financing, receivable accounts management and collection, and credit risk management.

The financial service sector of logistics solves many complex problems and difficulties relating to finance. The high capital turnover pressure that financing practitioners and logistics firms are faced with makes this crucial. Financially, it supports logistics and spot transactions, and increases fund turnover efficiency of both suppliers and consumers. Simultaneously, the stable and efficient development of a company's logistics financial business is guaranteed, thus a win for multiple parties.

4. Big data application service sector

The big data application service sector takes the Internet and corporate intranets as network support by applying technologies like cloud computing, data mining, geographic information systems, data warehouses, and business intelligence. This integrates all types of information from commercial, production, and logistical enterprise into a system database for comprehensive management. It aims to build a unified information exchange system, organize effective data exchange, and release real-time freight information. In addition, it integrates various application systems and aggregates all sorts of information to perform direct and visual analysis as well as predictions of the current business status. Thus, it offers data support for subsequent business development in logistics.

The big data application service sector includes statistical analysis, predictive analysis, operation optimization, and smart commerce. Below are more detailed descriptions:

1) Statistical analysis business

This is the harnessing of big data technology to summarize various participant information in logistics. The data is collected, filtered, and organized for the integrated and standardized management of basic business information. With big data application services, the logistics industry is able to lay a good foundation for follow-up business, so analysis can be rationally made to further consolidate and enhance customer relationships.

2) Predictive analysis business

Big data can help companies outline customer behavior and demand information. The logistics industry relies on data centers to process statistical data related to the smart platform. It extracts valuable information for statistical, predictive, and operational analysis, and reflects market demand changes with authentic and effective data. As a result, it predicts the development trend of core business, auxiliary business and value-added business, so as to predict each stage after the product is launched to market, reasonably control the inventory of logistics enterprises and make transportation plans.

3) Operation optimization business

Operation optimization business refers to the effective processing of data generated in logistics operations through applying big data technology. It quickly analyzes data correlations to make reasonable design arrangements and optimize transportation, warehousing, and storage. Consequently, it responds to new developments and makes arrangements as quickly as possible to ensure smooth industry progress. Ultimately, it improves the information level and predictability of enterprise to a large degree, thus their intensive distribution business management is smarter.

4) Smart commerce business

Smart commerce combines the reality of each logistics company, by applying modern data analysis methods, to transforming business data with high accuracy into information with commercial value. It also provides consulting and design services for integrated logistics. It can markedly improve the degree of system intelligence in a region as well as making enterprise management and control smart.

By carrying out these big data application service businesses, logistics effectively gathers and integrates resources both inside and outside the platform, such as, vehicles, goods, and warehouses. It also greatly alleviates the negative impact from the untimely transmission of information, and increases the frequency of transactions, the vehicle loading rate, warehouse utilization, etc. Meanwhile, the predictive analysis of data provides a basis for logistics companies to foresee and plan for future development trends.

3.1.2 Bulk commodity supply chain service business cluster

The bulk commodity supply chain revolves on core enterprise to connect suppliers, manufacturers, distributors, retailers, and end users into a unified functional network chain. This is achieved by controlling business, information, and fund flows from the procurement of raw materials to intermediates and final products, during which logistics companies transport bulk commodities to downstream businesses.

The supply chain, guided by customer needs, pursues better quality and higher efficiency. It integrates resources to achieve an organizational form where product design, procurement, production, sales, and service are efficient and collaborative. With the advancement of information technology, the supply chain has entered a new smart stage which is deeply entertwined with the Internet and IoT. To accelerate its innovation and application, industrial organizations, business models, and government ruling styles are innovated and supply-side structural reform is promoted.

To develop the bulk commodity supply chain, it is necessary to adopt modern management concepts and technology to reengineer business processes, perform intensive management and control, establish a scientific smart platform, render bulk commodity supply chain services, and upgrade informatization. This will contribute to long-term development and improve domestic supply chain enterprises.

The business cluster of bulk commodity supply chain service, as shown in Figure 3-3, is composed of supply chain management, finance, and collaboration, and big data service and consulting.

1. Supply chain management business sector
This is to coordinate the internal and external resources of an enterprise to meet customer demand. The enterprises in each link of the supply chain should be treated as a virtual enterprise alliance. Specifically, it involves cooperative enterprise management, and procurement, inventory, and sales services.

1) Cooperative enterprise management
The logistics industry manages enterprise cooperation in the bulk commodity supply chain. Through the collected information, it engages in the management business of production, trade, and logistics enterprises, and offers decision-making support (to them) based on basic information management. It also provides services for upstream and downstream cooperative enterprises, customer information management services for manufacturers, and business capability assessment services for all types of enterprise.

2) Procurement service
Based on the management of basic information, companies render procurement services for bulk commodity supply chains, realizing information exchange and resource sharing between upstream and downstream companies in the supply chain. They enable purchasers to quickly obtain

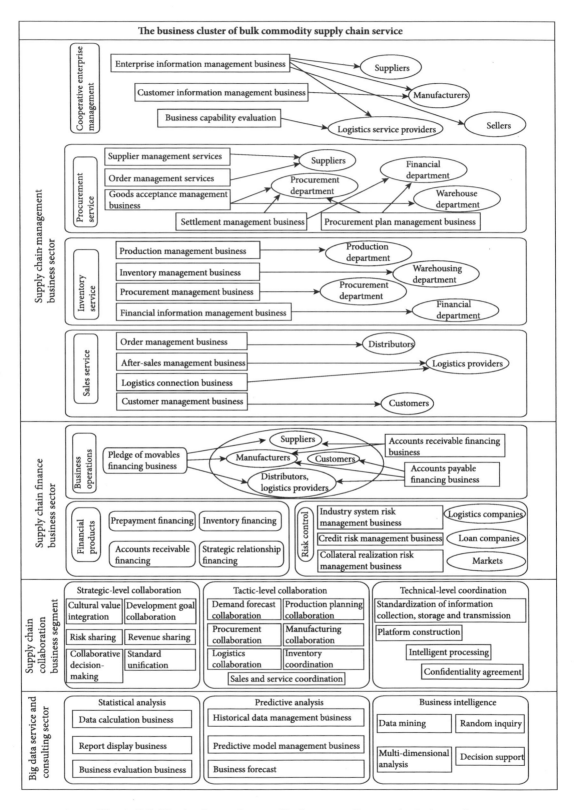

Figure 3-3 The business cluster of bulk commodity supply chain service

information via the release and sharing of public information and broaden the scope of potential suppliers. Specifically, it contains supplier, order, receipt, and procurement plan management, etc.

3) Inventory service

To solve excessive enterprise material reserves, excessive capital occupation, and low capital turnover, logistic enterprises have created inventory information services, which results in economic rationality and refined management through sharing information between connected enterprises at each node of the supply chain. Specifically, it comprises of production, inventory, procurement, and financial information management.

4) Sales service

To be fast, timely and efficient, every company in the logistics industry should manage sales with a tighter grip, thereby improving the operational level and efficiency of the supply chain. Targeted at the dealers, logistics providers and petroleum customers, coal power, and coal chemical industry, it should make management services available regarding customers, after-sales, and customer orders. This involves order management, after-sales management, logistics linkage, and customer management.

2. Supply chain finance business sector

Supply chain finance involves banks rendering finance, settlement and wealth management services to customers (core enterprises), as well as loan services to their suppliers, or prepayment and inventory financing services to their distributors. Specifically, it covers business operations, financial products, and risk control.

1) Business operations

In order to solve the financing difficulties of various companies in the bulk commodity supply chain, logistics companies render them with business marketing, logistics operations, and fund flow operation services. Financing solutions are given to promote effective interaction between finance and industrial economy.

2) Financial products

Based on the financial business needs of the supply chain, financial products are mostly targeted at corporate financing. This includes advance payments, inventorys, accounts receivable accounts, and strategic relationships. They can address issues concerning funding shortages during a purchase, liquidity, and company inventories. They are also used to deal with long-term funding shortages when downstream companies in the supply chain engage in credit sales, thus guaranteeing future capabilities in advance. Once a complete supply chain financial system is established, financial pressure is alleviated, costs are reduced, and fund security, transparency, and information is safeguarded.

3) Risk control

In response to the numerous financial links in the supply chain and involvement of many different enterprises, risk control is implemented. It establishes macro industrial prosperity indexes to predict the macro impact on business, issue early warning signals, and conduct emergency management of business risks. It also collects and analyzes the credit data of logistics and loan companies in the supply chain for real-time monitoring; analyzes and predicts market value and sale changes of commodities; grasps the real-time situation of collateral, and effectively avoids business losses. Specifically, there is industrial system risk, credit risk, and collateral realization risk.

3. Supply Chain Collaboration

Supply chain collaboration refers to a network consortium formed by two or more companies in the means of company agreements or joint organizations for a certain strategic purpose. On the one hand, this is to cope with increased competition and enhanced environmental dynamics; on the other hand, it builds a competitive advantage group and maintains this competitiveness at its core. Aimed at maximizing the benefits of each enterprise, every logistics company is supposed to develop supply chain collaborative business. Specifically, this includes strategic-level, tactic-level, and technical-level collaboration.

1) Strategic-level collaboration

Modern logistics companies have to clarify and strengthen the concept of collaborative supply chain management from a strategic perspective, further improve strategies and methods, enhance their own competitiveness in regional supply chains, and optimize collaborative solutions to supply chain issues. Specifically, this involves cultural value integration, development goal collaboration, risk sharing, revenue sharing, collaborative decision-making and standard unification.

2) Tactic-level collaboration

This is the center of supply chain collaboration management. Specifically, there is demand for collaborations in forecasts, production planning, procurement, manufacturing, logistics, inventory, and sales and service between upstream and downstream companies with a direct supply and demand relationship.

3) Technical-level collaboration

Modern logistic firms adopt collaborative technology to provide supply chain node companies with a real-time interactive sharing and communication platform for synchronous operation and information coordination. Transparency is improved, and so is the speed and effectiveness of decision-making. Specifically, information collection, storage and transmission, platform construction, intelligent processing, and confidentiality agreements are standardized.

4. Big data service and the consulting sector

1) Statistical analysis

Modern logistics companies, with network support from the Internet and corporate intranets, collect, filter, and sort out various data and information of trading, production, and logistics companies for the integrated and standardized management of basic business information. This lays a solid data foundation for subsequent business, and renders information display and release services based on the participating entities of the logistics business. Specifically, there are data calculation, report display, and operation evaluation businesses.

2) Predictive analysis

Based on the results of big data statistics on the national bulk commodity supply chain, modern companies can apply cloud computing, data mining, geographic information systems, and data warehouses to perform statistical, predictive, and operational analysis on supply chain service data. This helps forecast development trends of core, auxiliary and value-added businesses. Meanwhile, technical support is provided for various logistics companies in the supply chain – specifically, historical data management, predictive model management, and business predictions.

3) Business intelligence

Modern logistics companies combine modern data analysis and techniques to transform business data into information with commercial value. They render consulting and design services for various logistics entities, and improve supply chain system intelligence. Specifically, the business contains data mining, instant queries, multi-dimensional analysis, and auxiliary decision-making.

3.1.3 Bulk commodity trading business cluster

Bulk commodity trade refers to the bulk commodity wholesale market – also known as the spot trading market – engaged in commercial transactions. It helps companies establish closer cooperative relations with customers, suppliers and partners, and cultivate customer loyalty while increasing revenue. By improving the efficiency of order processing, the bulk commodity electronic trading platform manages commodity categories and catalogs – an action that requires the building of a complete catalog system for step-by-step evaluation. Regarding the transaction management process, market analysis enables process management and control in electronic trading, contract management, and order services; standardization is implemented in trade financing, commercial factoring, and loan management; and data warehouse and development technologies are applied for docking between platforms and users.

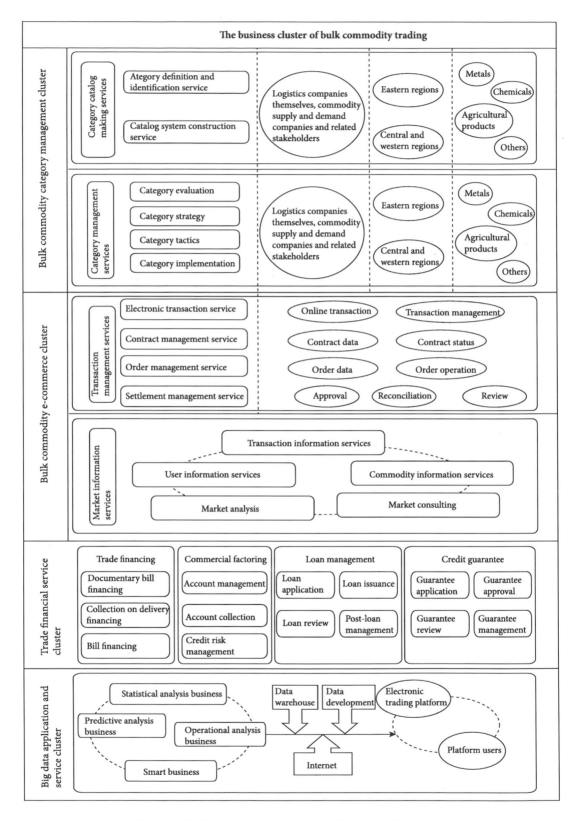

Figure 3-4 The business cluster of bulk commodity trading

The bulk commodity electronic trading platform is divided by nature and service sector. As shown in Figure 3-4, it consists of four business clusters: bulk commodity category management; bulk commodity e-commerce; trade financial service; and big data application and service.

1. Bulk commodity category management cluster

The making of a commodity category catalogue, targeted at all kinds of commodities in the warehouse, forms a unified and complete product category to facilitate the effective management of goods by collecting, counting, and sorting detailed information of the commodity itself, and of the production companies. Commodity quality review and certification, based on the category catalog of the trading platform, are performed upon the warehouse-stored goods. For example, checking whether the actual product is consistent with the manufacturer description and whether its quality meets industry standards. Commodity stock management emphasizes collecting the label information of commodities when they enter the virtual inventory; commodity inventory management enables inventory queries and stocktaking.

2. Bulk commodity e-commerce cluster

The bulk commodity e-commerce cluster embraces bulk commodity electronic transactions, transaction management, settlement management, and transaction review. As the core business cluster of the platform, it covers all electronic transactions as a one-stop bulk commodity electronic trading service.

3. Trade financial service cluster

Trade financial service plays the role of a value-added business cluster for the platform. It often centers on loans and advance payment, providing and publishing financial investment information, and warning of and controling financial risks.

4. Big data application and service cluster

Based on the massive commodity transaction data, big data application and service, via data mining, online analysis, and data combination and management, offers platform and decision-making support support for enterprise and platform users. This includes data warehouse management, data display, and decision-making assistance.

3.1.4 Logistics park service business cluster

1. An overview of logistics parks

Logistics parks are an inevitable outcome of modern logistics reaching a certain level of development. It is the staging ground of these companies for the purpose of intensification, common operation, and rationalization of the spatial layout of urban logistics facilities.

The first logistics park was built in Tokyo, Japan in the 1960s. It is called a logistics complex in Japan, a freight village in Germany, and a logistics base or center in China. In addition, the United

Kingdom, the United States, Belgium, Canada, and Mexico have also established logistics parks and freight distribution centers one after another. At present, the planning, construction and operation of logistics parks are in the ascendancy worldwide as a key facet of the contemporary industry.

According to the Fourth National Logistics Park (Base) Survey Report released in July 2015 by the China Federation of Logistics and Purchasing and the China Society of Logistics, there was a total of 1,210 logistics parks under planning, construction, and opeartion in China (as of 2015). Like the national Eleventh Five-Year Plan divides the country into eight major economic regions, these economic regions are also eight major logistics zones.

The Northeast Economic Region: Liaoning, Jilin, and Heilongjiang provinces

Northern Coastal Economic Region: Beijing, Tianjin, and Hebei and Shandong provinces

Eastern Coastal Region: Shanghai, and Jiangsu and Zhejiang provinces

Southern Coast: Fujian, Guangdong, and Hainan

Southwest: Yunnan, Guizhou, Sichuan, Chongqing City, and Guangxi Zhuang Autonomous Region

Northwest: Gansu, Qinghai, and the autonomous regions of Ningxia Hui, Tibet, and Xinjiang

Central Region: Shaanxi, Shanxi, and Henan provinces, and Inner Mongolia Autonomous Region.

Yangtze River Region: Hubei, Hunan, Jiangxi, and Anhui. The distribution of logistics parks is shown in the following table.

An overview of the distribution of logistics parks in each economic region across the country in 2015

Region	Under planning	Under construction	Under operation	Total
The northern Coastal Economic Region	5	30	181	216
The economic region in the middle reaches of the Yangtze River	35	50	126	211
The economic region in the middle reaches of the Yellow River	21	40	114	175
The eastern coastal economic Region	9	14	133	156
The southern coastal economic Region	10	18	107	135
The southwestern economic region	16	38	78	132
The Northeast Economic Region	11	32	68	111
The northwestern economic region	6	18	50	74
Total	113	240	857	1210

2. Division of service business clusters in logistics parks

The rapid development of modern logistics parks, continuous scale expansion, and their gradually increasing market share have enabled us to consider expanding these services. They can adapt to the new needs of urbanization, industrial transformation and upgrade, and internationalization, for

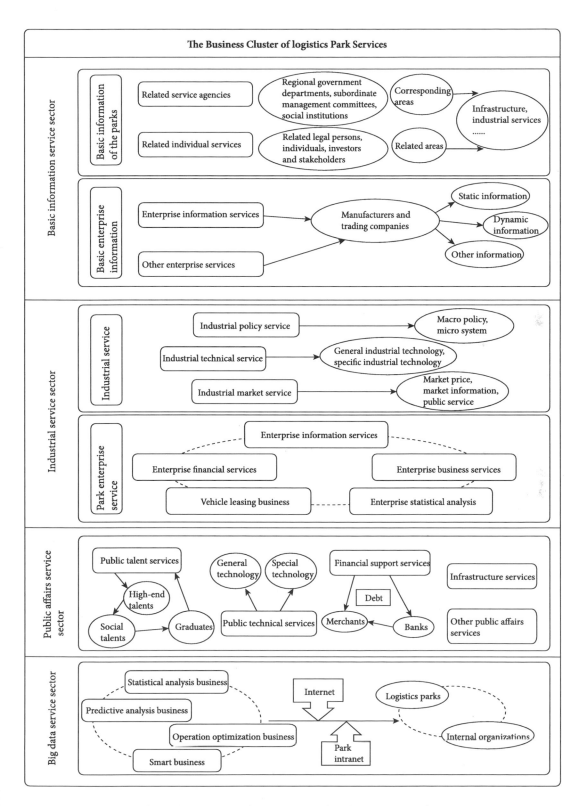

Figure 3-5　The business cluster of logistics park services

optimum benefit. As shown in Figure 3-5, the service business cluster should take the construction of the information service platform as the focal point and develop basic information, industry, public affairs, and big data services.

1) Basic information service

This mainly embraces the basic information of the park and of enterprise, service and management of the relevant entity information. The basic affairs service, as the basis for the logistics park to complete various services and management tasks, can record, inqure, modify, and add basic information of the park, related institutions, departments, enterprise, and individuals, and assist the development of economic and public affairs. Ultimately, it provides information support for the development of industry and enterprise in the park.

2) Industrial Service

The industrial service cluster mainly includes enterprise and industrial linkage services. With the goal of assisting economic and industrial development, it forms the cornerstone for industrial upgrade, serving both industry and enterprise. As a result, it promotes the informatization and integrated development of the park economy and has a huge developmental impact.

3) Public affairs service

With park managers as the carrier, this mainly includes the park's public services and e-government service. As the basis of the park's information management and industrial development, it covers all aspects of its public affairs, such as public talent, financial support, technical, infrastructure, and energy conservation and environmental protection services.

4) Big data service

The big data service, based on massive data, employs big data technology to provide support for the whole logistics park service business via data collection, data processing, etc. It mainly involves the management of basic data, demand forecasting, and data support for smart business, operational analysis, and statistical analysis.

3.2 The business system design of the smart platform

The design of the smart logistics and supply chain information platform should be based on the four core businesses of modern logistics to build a complete system structure underpinned by relevant information technology. Specifically, the business system is designed for the smart logistics, bulk commodity supply chain information, bulk commodity electronic trading, and logistics park information platforms.

3.2.1 The business system design of the smart logistics information platform

Smart logistics, centered on the Internet +, receives technical support from the IoT, cloud computing, big data, and the three networks (sensor network, IoT, and Internet). Based on automation infrastructure, intelligent business operation, information system-assisted decision-making, and key supporting resources, it is a form of logistics where the entire process chain can be automatically sensed and identified, traced, instantly responded to, and intelligently decided. Optimization occurs by seamlessly integrating the information systems of various logistics links and enterprises.

This platform can realize the intelligence and integration of logistics business operations by using information integration technology. Its planning and design aim to integrate logistics services, visualize processes, electronicize transactions, intensify resources, standardize operations, and personalize customer service. Combined with the regional logistics ecology and specific business needs of enterprise, they strive to determine the corresponding operation subjects, service scope,

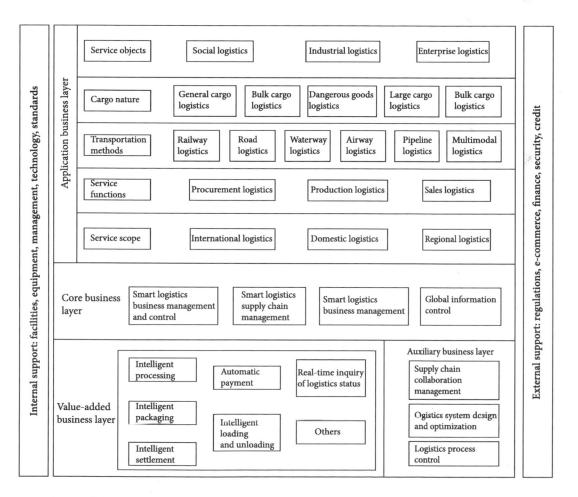

Figure 3-6   The business system of the smart logistics information platform

business system, and input and output. This ensure a complete set of functions and a reasonable system on the platform.

Modern logistics relies on the road, rail, and waterways to establish and gradually improve its transportation network. On top of that, it provides customers with a complete set of solutions that integrate transportation, finance, and information management. With the analysis of industry clusters, market positioning, and development strategies as well as the combination of design concepts of the platform, existing businesses are sorted and rectified. By operating attributes, they have three categories: core business; auxiliary business; and value-added business. Based on this, the business system of the smart platform is designed as shown in Figure 3-6.

It can be seen from the diagram above that the smart platform accepts support from both internal environments (such as facilities, equipment, management, technology, standards) and external environments (such as e-commerce, finance, security, and credit). Targeted at distinct commodity types and service subjects, it carries out a series of core business objectives, which includes smart business management and control, smart logistics supply chain management, and auxiliary business. These three major businesses together constitute the smart logistics information platform.

3.2.2 The business system design of the bulk commodity supply chain information service platform

The bulk commodity supply chain information service platform, built on the idea of supply chain management coordination, matches with the business needs of upper and lower entities in the supply chain. This allows the operation process and information system to be closely coordinated and all links seamlessly connected, forming the lead model of logistics integration, and information, document, business, and fund flows. Consequently, it enables the visualization of the overall supply chain, informatizaion of management, maximization of overall benefits, and minimization of management costs.

Guided by the Belt and Road Initiative, the logistics industry should make full use of resource, location and industrial advantages when relying on the four major businesses of the bulk commodity supply chain service. Then it can build a supply chain management platform that conforms to modern logistics and renders production, trade, and related logistics companies with major supply chains.

Based on the functional positioning of the service platform, detailed analysis and sorting of industrial chain development leads to its classification by business attribute. This includes core, auxiliary, and value-added businesses. This is shown in Figure 3-7.

The diagram depicts the supply chain management of logistics, as supported by the internal environment (facilities, equipment, management, technology, standards) and external environment (regulations, e-commerce, finance, security, and credit). It targets different service subjects and industries to engage in a series of core, auxiliary and value-added business activities.

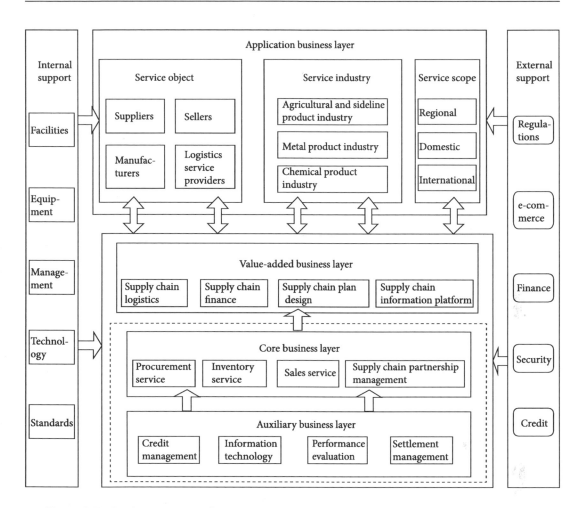

Figure 3-7   Business system of the bulk commodity supply chain information service platform

1. Core business

The core business layer contains procurement, inventory, and sales services as well as supply chain partnership management. And bulk commodities are roughly divided into three categories: agricultural and sideline products, metals, and chemicals. Most of the relevant logistics companies in these industries lack a unified supply chain management system and guideline. Therefore, they require core business services.

2. Auxiliary business

The auxiliary business layer covers four main businesses: credit management, information technology, performance evaluation, and settlement management. The business system of the modern logistics supply chain management, as a unified architecture of the bulk commodity supply chain, adopts information technologies such as the IoT, big data, and cloud computing. This is done for unified credit management, performance evaluation, and settlement management for supply chain node companies to nurture the healthy development of the industrial chain.

3. Value-added business

The value-added business layer includes supply chain logistics, finance, plan designs, and information platforms. Based on the core and auxiliary businesses, supply chain management value-added business takes place, openning up new sources of profit. This includes logistics services provisions between supply chain node enterprises. Thus, financial services are rendered to relieve capital pressure and reduce financing costs, while a supply chain information platform is built.

4. Application business

These supply chain management businesses primarily serve four types of node enterprise in the bulk commodity supply chain: suppliers, manufacturers, sellers, and logistics service providers. They also serve three major industries in the bulk supply chain: agricultural and sideline products, metals, and chemicals. As it further expands, it will reach more fields.

5. Internal and external support

As the basis for the construction of the supply chain management business system, internal and external support projects ensure the smooth execution of core, auxiliary and value-added business. The internal support concerns management facilities and equipment, and technology of the supply chain; external support refers to the regulations of bulk commodity supply chain management – e-commerce, finance, security and credit.

3.2.3 The business system design of the bulk commodity electronic trading platform

The establishment of this trading platform enables the rapid turn over of bulk commodities. Engaging in bulk commodity electronic trading gradually integrates bulk commodity resources and provides electronic transaction services for production and sales companies and trade demand companies, thereby improving the trading environment, simplifying the transaction process, and enhancing transaction efficiency. Based on the platform's functional positioning, detailed analysis and sorting of related businesses lead to the clarification of business operation attributes. This is shown in Figure 3-8.

Figure 3-8 points out that the business system of the platform, supported by the internal environment and regulations, engages in a series of core, auxiliary and value-added businesses. The core business consists of online trading, transaction management, settlement management, market information, and market analysis; auxiliary business management embraces commodity information, commodity display, inventories, and corporate information; and value-added business has five components: loans, advance payments, investment and financing, membership transactions, and big data service. These three business layers constitute the bulk commodity electronic transaction business system.

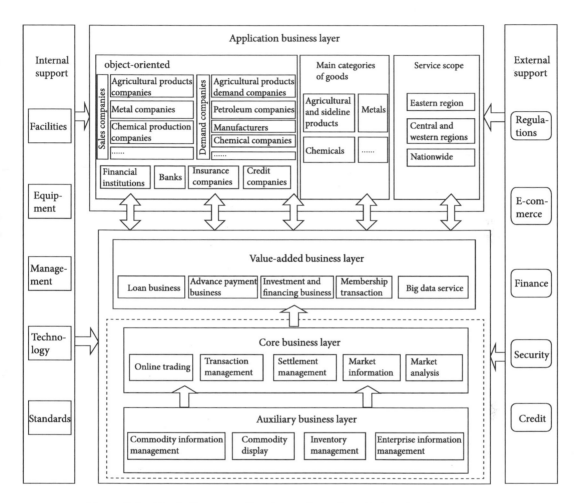

Figure 3-8   The bulk commodity electronic trading platform business system

3.2.4 Design of the logistics park information service platform

Based on the current economic development of modern logistics parks, this information service platform can provide a complete set of service solutions that integrate public, industrial, corporate, and big data application services. This ranges from its target planning and functional positioning to the park and corporate customers in it. And the analysis of industrial clusters, market positioning and development strategies, and sorting of existing businesses lead to operational attribute classification. This includes basic, core, public, and big data application service businesses. This is presented in Figure 3-9.

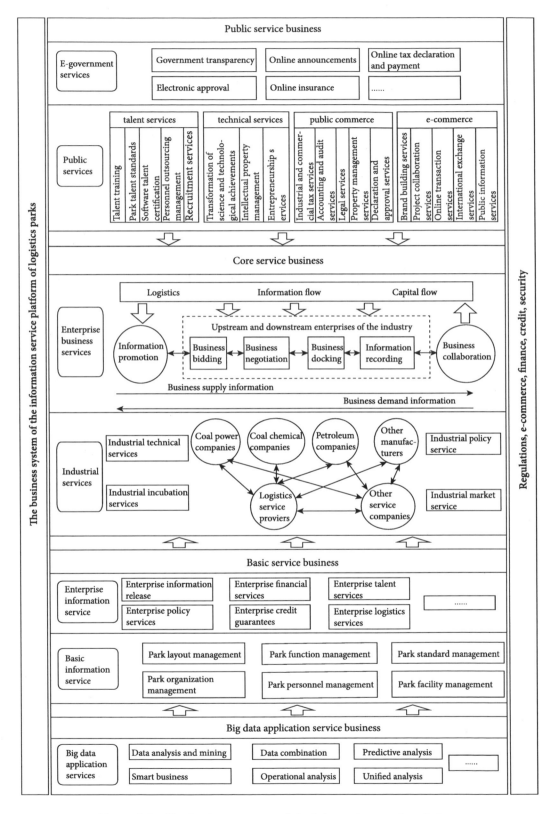

Figure 3-9 The business system of the information service platform

Figure 3-9 elaborates on the business system of the information service platform. It takes support from external environments such as regulations, e-commerce, finance, credit, and security. It also engages in a series of service businesses with different parks and enterprise. Basic service business includes basic park information and enterprise information; core service consists of industrial park and park enterprise business services; and public service business is compromises of public and e-government services. The four key businesses together constitute the business system of the logistics park.

CHAPTER 4

DEVELOPMENT STRATEGIES AND MODEL INNOVATION OF THE INFORMATION PLATFORM

The effective and correct formulation of the smart information service platform should be based on the positioning and development objectives of the logistics industry service platform. It should combine development status with actual demand in the supply chain, and consider medium- and long-term forecasts. Specifically, it provides strategies on intensive integration services, supply chain integration, industry convergence extension, and collaborative development.

4.1 Connectivity between national initiatives, strategies, and the informatization of logistics

4.1.1 The Belt and Road Initiative facilitates the informatization of the logistics industry

The Belt and Road Initiative is an abbreviation – the 'Belt' is short for Silk Road Economic Belt, and 'Road' for the 21st Century Maritime Silk Road. It advocates peaceful development, actively building economic partnerships with countries along the route. As the initiative deepens, trade between China and these other nations has grown closer. Import or export, logistics is absolutely indispensable. Meanwhile, the demand for informatization is increasing, which pushes forward adjustments and accelerates full informatization in the industry.

4.1.2 The Internet + accelerates informatization

Internet + represents a new economic form. It refers to the integration of the Internet and traditional industries. This is supported by Internet information technology to accomplish economic transformation and upgrade by optimizing production factors, updating business systems, and reconstructing business models. The purpose of the Internet + plan is to make full use of the Internet, deepen integration, and promote economic productivity. The ever increasing scale and efficiency of logistics leads to the inevitability of informatization. Under the background of Internet +, it is greatly beneficial to accelerate informatization in logistics. This can achieve informatization upgrade, enhance competitiveness, and expand competitive advantage.

4.1.3 *Made in China 2025* propels informatization

Made in China 2025 covers the first ten years of China's manufacturing power strategy. It aims to accelerate the transition from Made in China to Intelligent Manufacturing in China, the country's fourth industrial revolution. For logistics, it is undoubtedly a new opportunity. In the past, China had a low-cost competitive edge. However, as labor costs keep rising, economic development has slowed, and the economy faces higher costs. This has pressured enterprise like never before, but it is also a powerful driving force for upgrading the traditional logistics industry.

4.1.4 Beijing-Tianjin-Hebei Integration improves informatization

The Beijing-Tianjin-Hebei region has enormous potential for economic development as well as having the densest logistics network. An efficient and mature system is not only a booster for the regional economy, but also an effective means and prerequisite to guide coordinated regional development. At present, its low logistics information level and the disunified information platform seriously hinder its development. With the implementation of the Beijing-Tianjin-Hebei integration strategy, the continuous improvement of informatization region will galvanize collaborative development. Therefore, accelerating the informatization of the logistics industry to create a smart and unified logistics system is crucial.

4.2 Strategic model and innovation of the smart service platform

The construction of the smart logistics and supply chain information service platform should take actual business operation and future expansion needs as the entry point, combined with the development status of logistics information platforms. It aims to accomplish supervision informatization, supply chain resource integration and service, electronic bulk commodity trade, and platform operation of industrial parks. This is not only feasible, but also innovative and advanced. Therefore, it must fully consider the development of the industry, its characteristics, and its medium and long-term needs, in order to propel it further.

The strategic system of the smart platform embraces an intensive and integrated service, supply chain integration, industry convergence, and collaborative development. Its implementation should make full use of existing network, logistics, and supply chain resources to develop advanced management ideas and modern network technology. This can help build a supply chain centered on core business and integrated resources, catering to market demand, expanding business scope, and establishing core market competitiveness. This is shown in Figure 4-1.

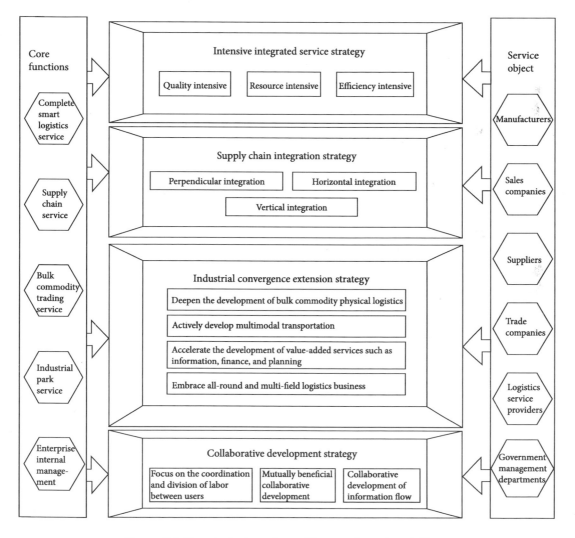

Figure 4-1 The strategic system of the smart service platform

4.3 Intensive integrated service strategy

The intensive development model of the smart platform integrates social resources through the big data information platform. Smart logistics, supply chain management, electronic bulk commodity trading, and park management are gathered on the platform to generate the agglomeration effect; this in turn attracts external logistics, and business, capital, and information flows to form an all-encompassing information service platform. The intensive integrated service strategy is illustrated in Figure 4-2.

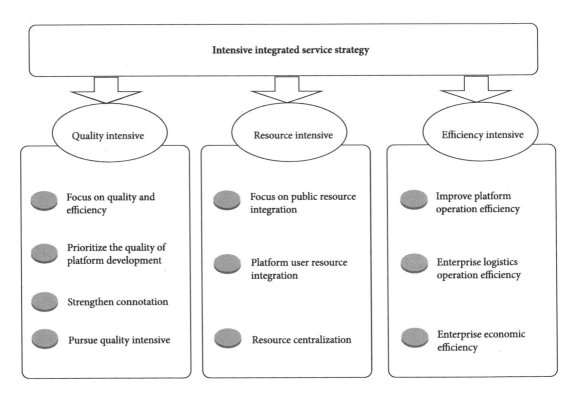

Figure 4-2 Intensive integrated service strategy

4.3.1 Quality-intensive development strategy

Intensive development revolves around quality and efficiency; it increases logistic industry interest. Based on a full understanding of the intensive development model, logistics should prioritize the development of the smart service platform. The business philosophy should shift the focus from extension and expansion to strengthening the core, pursuing asset quality, improvement, platform management quality, and platform service quality for optimum outcome.

4.3.2 Resource-intensive development strategy

As a regional public service platform, the smart service platform should follow the integration of public logistics and other resources, and change the business ideas of decentralized and isolated operation by region. All kinds of logistics, business, capital, and information flow resources of platform users are systematically integrated, forming an intensive development model with centralized and large-scale resources.

4.3.3 Efficiency-intensive development strategy

The logistics industry should emphasize the optimal allocation and recombination of limited resources. With information technology as the primary technical support, it increases work efficiency, improves operating efficiency of the smart platform, and obtains higher operating income. The objective of developing this platform should be to achieve higher efficiency.

4.4 Supply chain integration strategy

The ongoing competition between logistics companies concerns the supply chain. A high level of supply chain management not only saves costs, optimizes the capital chain, and spares more liquid funds for logistics companies, but improves customer response times and service quality. The construction of the smart platform should adopt the supply chain development strategy.

This integrated information platform consists of four platforms: smart logistics, bulk commodity supply chain service, bulk commodity electronic trading, and industrial park services. It should fully analyze the characteristics and needs of each supply chain and build in-depth business cooperation with upstream production companies, downstream end customers, intermediate business enterprise, logistics companies, and finance and other public service institutions. This integration strategy is demonstrated in Figure 4-3.

The diagram displays the integrated management of the four platforms which concern business, capital, and information flow, and the supply chain integration strategy design from three dimensions: perpendicular, vertical, and horizontal.

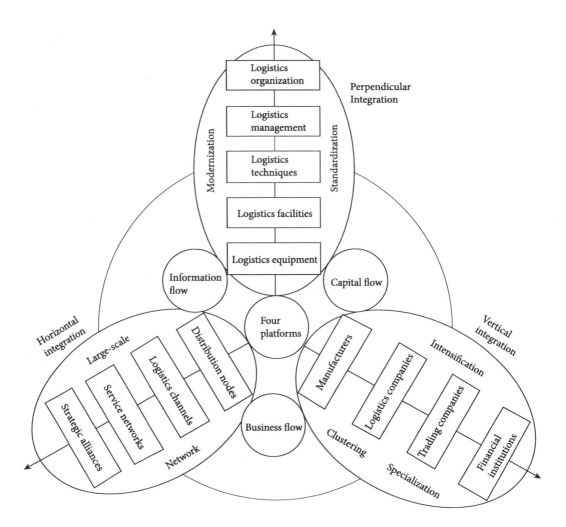

Figure 4-3 Supply chain integration strategy

4.4.1 Perpendicular integration

The perpendicular integration of the supply chain is to make the organization, management, technology, facilities and equipment of logistics consistent with the development direction of modern supply chain. This is in line with the actual business needs of the specific supply chains in which modern logistics companies are located. This strategy requires supporting policies, material resources and economics. It promotes the development of modern enterprise by establishing an effective organization and management system, and implementing an effective management mechanism.

4.4.2 Vertical integration

Vertical integration of the supply chain effectively integrates production, logistics, and trade companies, financial institutions, and government departments to allay the bullwhip effect. This establishes a dynamic and rapid response mechanism for the supply chain, saves management costs, and improves operating efficiency and interests. It thereby accomplishes the professional and large-scale development of enterprise and institutions at each supply chain node.

4.4.3 Horizontal integration

Horizontal integration mainly embraces the establishment of collection and distribution nodes, the construction of logistics channels, and the coverage of service networks. In this environment, resource allocation efficiency, and management and operation will be greatly improved, while costs and risks drop drastically. It gives full play to the powerful agglomeration effect of the four platforms, strengthens cooperation of strategic supply chain alliances, and engages in integrated operations for mutual benefit.

4.5 Industry convergence extension strategy

Removing the drawbacks of singular operation, while strengthening industrial integration and upgrade as well as broadening the core business, are effective ways to transform modern logistics. They should carry out industrial expansion along all dimensions of the value chain, and exercise scientific, moderate, and efficient judgement, via vertical integration and horizontal extension. The industrial convergence and extension strategy is shown in Figure 4-4.

This strategy is divided into four levels. Expansion can either occur simultaneously or step-by-step.

4.5.1 Deepen the development of physical logistics of bulk commodities

Modern logistics companies ought to combine their own superior resources, starting with physical logistics. They can play the leading role of the four major information platforms, and deepen the establishment of long-term cooperative relationships with core companies. This broadens business scope and increases profit margins.

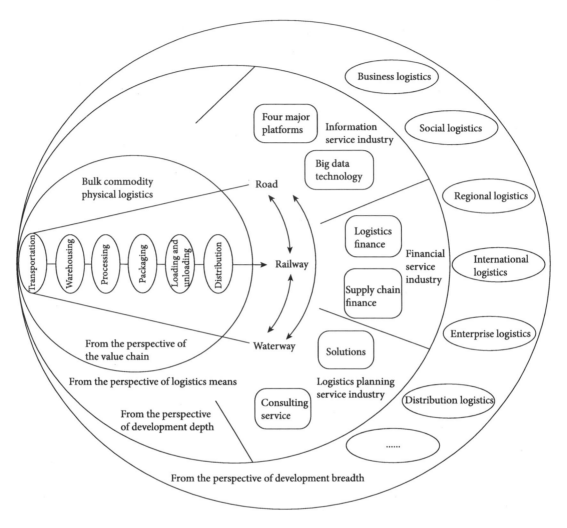

Figure 4-4 Industrial Convergence Extension Strategy

4.5.2 Actively develop integrated multimodal transportation

Different logistics companies may have an absolute advantage in a certain mode of transportation, but as customers always require better and better services, a single mode of transportation rarely suffices, nor is it economical. Therefore, modern firms should utilize their own merits to enrich transportation methods and organically combine different methods.

4.5.3 Accelerate the development of value-added services

As modern logistics keeps marching forward, more extension services have emerged, such as financial, information and consulting. Modern logistics companies should combine the four major info platforms to launch extension services, such as professional information service companies and supply chain financial centers. This lays a solid industry foundation and broadens the services on offer.

4.5.4 Embrace the all-round logistics business

Modern logistics companies should not limit their business to a certain field, but boldly engage in international logistics (freight forwarding, customs clearance and commodity inspection), and commercial, social, and regional logistics. This can significantly improve certain companies involved.

4.6 Collaborative development strategy

Collaborative development is needed to coordinate the resources, social conditions, economy and environment of the logistics companies, enabling them to apply the four information platforms. The application, centered on the information platform of logistics companies, relies on commodity trading and industrial park service platforms to strengthen cooperation and make mutual progress; ultimately it promotes the collaborative development of the logistics industry. This is shown in Figure 4-5.

4.6.1 Coordination and division of labor between users

The basis for collaborative development is the economic connection between platform users. User association and division of labor are the theoretical and economic foundations for this. Companies provide a big data information platform, mostly used by commodity supply and demand companies. The platform releases information about the source of goods and its flow direction. And third-party financial institutions improve trade finance and insurance business in accordance with this. The platform users are interdependent, forming a value-added process through the flow of materials, funds, and information.

4.6.2 Mutually beneficial collaborative development

Collaborative development, which breaks boundaries between modern logistics companies and regions, is a two-way or multi-directional behavior of mutual benefit. From the perspective of modern logistics firms, it promotes mutual establishment of strategic alliance between enterprise and users. Also, linkage increases the degree of information shared between companies, thereby reducing transaction risks. From a regional perspective, collaborative development accelerates regional industrial expansion.

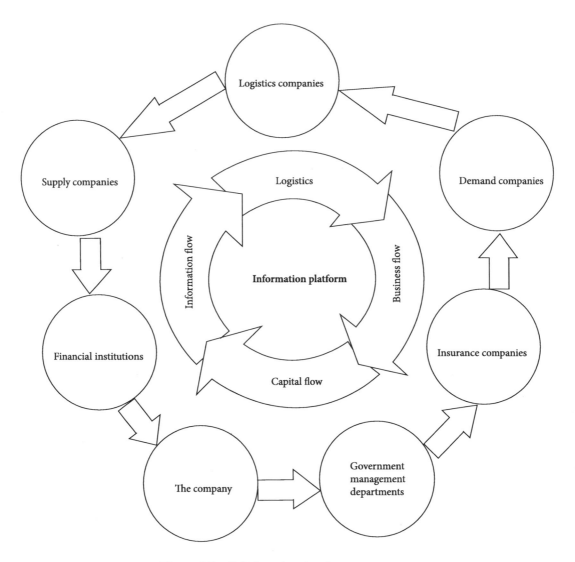

Figure 4-5   Collaborative development strategy

4.6.3 Collaborative development of information flow

Modern logistics companies should make use of the smart platform to integrate various types of information to form a highly concentrated information flow. The sharing of information is enabled by exchange with platform users, thus improving operating efficiency, and reducing operating costs.

4.7 Research on the construction mode and innovation of the smart platform

4.7.1 The construction mode

According to the overall plan and strategic deployment (and characteristics of the logistics industry), the functions of the four major information platforms are analyzed, and the construction modes and innovation points of each platform are summarized. To ensure the smooth development, construction, and operation of the smart service platform, a comprehensive construction mode of government guidance plus enterprise construction is adopted, as shown in Figure 4-6.

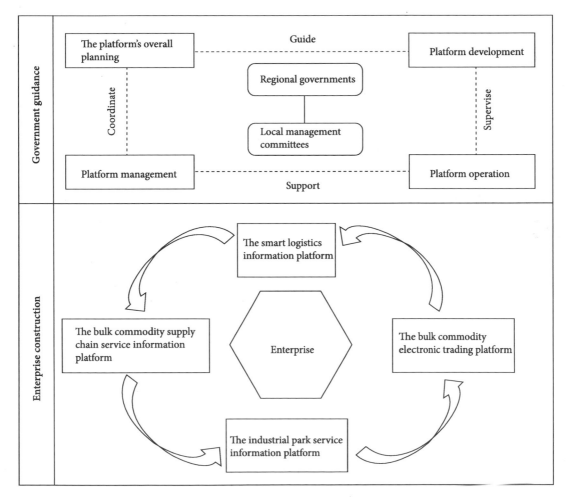

Figure 4-6   The construction mode of the smart service platform

1) Government guidance

Based on the construction of the smart platform, regional government unites relevant industry alliances, associations, and technology service companies, forming a management committee and logistics productivity promotion center. They will jointly guide and promote the overall plan and design of the platform, coordinate relevant development, assist in problem solving, and continuously improve regional logistics environments. This is achieved through policies and regulations, business procedures, and technical standards. It makes full use of the productivity promotion center to improve market-oriented operation of the information platform, and enhances platform service quality and operational efficiency. The management committee plays the role of organizer, manager, and supervisor, guiding, supervising, supporting and coordinating platform development. During its operation and management, the committee always offers corresponding guidance and supervision.

2) Enterprise construction

In the actual development of the platform, modern logistics companies are responsible for related construction work. Guided by the management committee, relevant development, management, operation, coordination, and maintenance take place in accordance with overall planning principles. This guarantees the smooth functioning of the four platforms. Upon completion, the construction company shall act as the main operating body in the running of the platform while it attract logistics companies, production and trade companies to adopt it. During the process, the construction company shall keep improving platform functions, innovate transaction modes, and integrate business, logistics, capital, and information flows. This helps build a cutting-edge logistics information platform.

4.7.2 Innovation

1) Platform integration innovation

The design and planning of the smart platform should begin with the actual needs and development trends of modern logistics companies; it should fully absorb the advanced experience of excellent domestic and foreign logistics information platforms. The concept of information system integration will penetrate the platform design to intensify and unify platform hardware, software, and data information. Thus, it is advanced, innovative and feasible. Consequently, it can render logistics informatization with a set of practical service solutions enabling various business and management departments to operate in a coordinated, unified, and efficient manner.

2) Platform construction mode innovation

To ensure the orderly development of the information platform, platform construction adopts the bottom-up collaborative construction mode of government guidance plus enterprise construction, which is gradually implemented in stages. This mode puts the construction company in charge of platform development, takes other related companies as its users, and gradually integrates regional logistics and market resources to build a complete regional logistics network; the government mostly guides, supports, supervises, and enhances the soft environment (such as regional logistics policies and regulations, and related standards to safeguard barrier-free communication of logistics big data), thereby improving operational efficiency and promoting the collaborative development of regional industries.

3) Platform operation mode innovation

As guided by the government, modern logistics companies create a community of interests centered on the government, logistics companies, and production and trade enterprise. This is done in accordance with the market-driven logistics informatization development mode. They integrate industry advantage and market experience of each company into a joint force, thus improving the operation and management level of the smart platform, its market position, market competitiveness, and reducing its operating risks.

CHAPTER 5

OVERALL STRUCTURE AND FUNCTION DESIGN OF THE INFORMATION PLATFORM

5.1 Guiding ideology and design philosophy

5.1.1 Guiding ideology

Based on its overall design goals combined with modern logistics industry clusters, business systems, and operation analysis, the guiding ideology of designing the smart logistics platform is to focus on the development of supply chain services. This can make use of the Belt and Road Initiative to build front-end resources and back-end market advantages for customers for an optimized configuration of international and domestic markets. With the help of Internet + thinking, the building of smart information and bulk commodity supply chain service platforms, to adopt logistics modularization, trade networking, service informatization accomplishes the built-in and seamless development of core business enterprise and customer business processes. This is supported by business and management mode innovation to utilize resources and service customization in the supply chain and value-added services.

5.1.2 Design philosophy

Based on the planning and design of the logistics platform combined with its characteristics being the core of supply chain logistics, the design philosophy revolves around the construction needs of the four industrial clusters – smart logistics; supply chain management; bulk commodity trade; and logistics park services. Thus, the building of the smart platform is a normative, advanced, expandable, open, reliable, cooperative and economic outcome, as shown below.

1. Normative

The smart logistics platform must support all open standards. Whether it is the operating system, database management system, development tools, application development platform, or workstations, servers and networks, all must comply with national, industrial, and computer software and hardware standards for global sharing of data and information and better data reuse.

2. Advanced

When constructing the smart logistics and supply chain information service platform, we shall plan IP networks carefully, and use advanced technical means and network products as much as possible to enable good performance and adapt to the rapid advancement of information technology, thus reserving space for future expansion.

3. Expandable

To maintain a long lifecycle, the logistics information platform must have fine expandability and environmental adaptability. It should heed the impact of future development of new technologies. By adopting a modular structure to improve the independence of data and program modules, it enables the system to have greater openness and structural variability, ensures convenience for platform transformation and upgrades, and improves the system's adaptability to new technology and applications.

4. Open

As an open system, the smart platform has to be connected to other external platforms or systems through interfaces for information exchange, which deserves thorough consideration during its planning and design.

5. Reliable

The business system of the information platform is directly oriented at users. The information flowing on the business system is directly related to their economic interests and there is a high degree of sharing between the systems. Therefore, it is essential to ensure the security of information transmission. The information platform must be highly reliable regarding security and confidentiality, error detection and correction, anti-virus capability, etc.

6. Cooperative

The smart service platform is composed of smart logistics information, a bulk commodity supply chain platform, bulk commodity electronic trading platforms, and a logistics park information service platform. It is required to integrate information from four platforms and multiple departments, thereby requiring participation from governments, enterprise, and information system developers in its development, maintenance and use. All parties involved are required to unify rules, cooperate, and participate actively for satisfying benefits.

7. Economic

On the premise of meeting information needs, the smart service platform should select the equipment with a high performance-price ratio to construct the information platform with high goals, low costs, and superior quality.

5.2 Overall technical route

Based on analysis of the business system and business process of the general logistics industry as well as the construction foundation and demand analysis of the service platform, overall design of the platform takes place under the premise of its strategic analysis. This includes logical structure design, network topology structure design, and key technology selection, which involves software and hardware configuration of system demonstration projects, platform system testing, and comprehensive evaluation. The overall technical route is shown in Figure 5-1.

5.3 Overall structure

Based on the investigation and analysis of the status quo of logistics industry informatization, the overall structure of the information service platform is achieved with guidance from the development strategy combined with its needs. The structure contains basic environment, IoT integrated application, application support, business application, and smart decision support layers. The overall structure of this platform is shown in Figure 5-2.

5.3.1 Basic environment layer

1. The platform-supported environment layer

The platform-supported environment layer, based on IoT, embraces the operation environment of the system and database warehouse environment, which support the operation of logistics and the supply chain, the use of development tools, web services, and large-scale data collection and storage, thus safeguarding the integrity of the whole platform's operating environment.

2. Network platform

The network platform mainly includes IoT, wide area, local area, mobile communication, and industrial private networks. This and related system interfaces can lay the foundation of web services and resource addressing to support information transmission of related businesses.

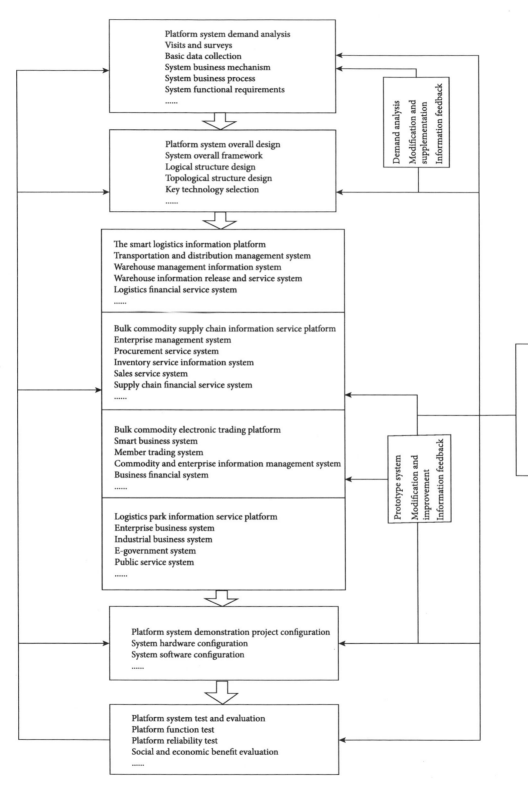

Figure 5-1 Overall technical route

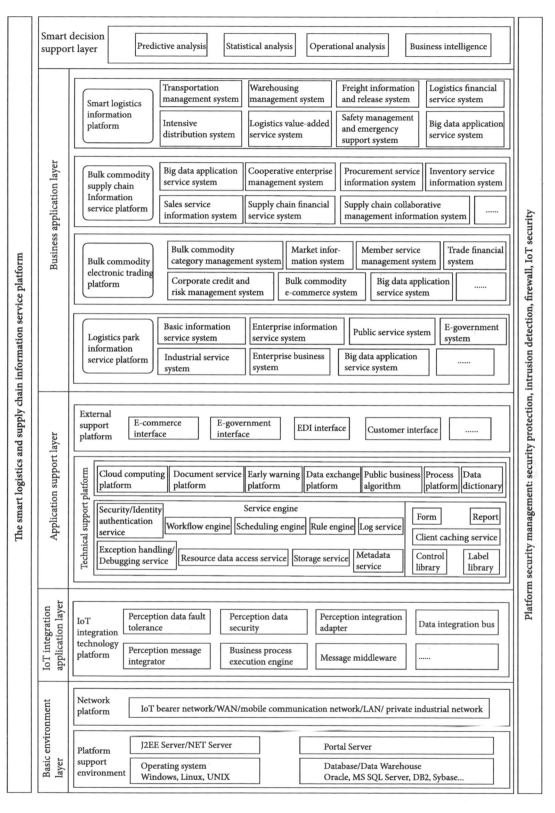

Figure 5-2   Overall structure of the smart logistics and supply chain information service platform

5.3.2 IoT integration application layer

The IoT integration application layer is the core component of the information collection function. Via the relevant information and data integration processing modules of the perception tool, it manages the data format required for message queue services, information management, data management centers, data exchange, and application integration in a unified manner.

1. Perception integration adapter

The metadata adapter converses RFID private data formats to metadata; provides the model library of registered RFID data to the enterprise-level and that of registered specific metadata to the transportation department-level, the data read-and-write adapter, RFID data read-and-write interface, and the data aggregation function during data access. The publishing/subscribing adapter provides a topic-based publishing/subscribing interface. The publishing/receiving adapter provides queue-based sending/receiving interface. The process adapter provides process information such as work lists and items from the business execution engine.

2. Data integration bus

Metadata management performs unification of data format definitions required for data management centers, data exchange and application integration; interface services provide interface standards between applications, converse different interface standards, and search within the companies, such as by communicating with ONS via the web service interface to find information and business functions available on the IoT.

3. Business process execution engine

The event-driven business process execution engine can receive multiple ALEs (Application Level Event) and merge them into the workflow to drive the execution and management of logistics and cross-system business processes.

4. Perceived message integrator

According to the content of one or more fields of the message header or body, the rules are customized to convey messages to different destinations through routing; to convert them so they can support a variety of formats including XML (extensible markup language) and flat file format; to modify the content of messages, such as adding data extracted from the integration bus to the message body.

5. Message middleware

It renders message queue services, which is the basis for asynchronous data exchange; seamlessly integrates ALEs between the edge layer and integration layer; manages information storage, including temporary storage for information exchange; converses rule-based information, which exchange various data within the logistics departments automatically in line with exchange rules.

6. Perceived data fault tolerance

When the perceptual system fails, the failure is automatically detected, the system is resumed to normal operation, or the program will not be suspended or modified because of it.

7. Perceived data security

This module safeguards data privacy and security through data encryption of the perceptual system, information authentication, and RFID tag blocking programs.

5.3.3 Application support layer

1. Technical support platform

On the one hand, the technical support platform combines service engines and resources, data access services and perception technologies based on key functions to perform data management and business process execution engine functions; on the other hand, it supports transmission, processing, conversion of perception data into business applications via cloud computing platforms, data exchange, data dictionaries, etc. In addition, it also introduces a set of related development tools and provides a professional, safe, efficient and reliable development tool platform that develops, deploys and operates logistics management application software for various complex commercial application systems.

2. External support platform

The external support platform is mainly the external interface required to complete various businesses. The logistics information platform based on IoT connects with the information systems of government and service agencies through e-commerce, customers, e-government, and EDI (electronic data interchange), so as to achieve collaborative work and services and effective information collaboration

5.3.4 Business application layer

The business application layer is an important window for the smart platform to render application services to end users. It aims to support their networking application requirements. As user requirements vary, the business application layer contains a variety of application entities that support different applications with distinct services. And the information service platform embraces four sub-platforms: smart logistics information, bulk commodity supply chain information, bulk commodity electronic trading, and logistics park information.

1. Smart logistics information platform

This adopts advanced system design, integration, and other information technology, transportation management, warehousing supervision, intensive distribution, freight information service and release, financial logistic service, logistics value-added service system, safety management and emergency support systems. This solves expansion, operation and service guarantee problems of the core business of transportation and logistics, integrate logistics resources, and provide customers with a complete set of logistics solutions that integrate transportation and information management.

1) Transportation management system

This is a comprehensive system that integrates transportation plan management, vehicle scheduling, dynamic real-time tracking, vehicle status and safety, order management, transportation volume, and cost statistics. It adopts modern information technology to effectively manage transportation planning, means, personnel and process tracking, dispatch and command, transportation tools and transportation process, which is conducive to improving its service quality.

2) Warehousing supervision system

A comprehensive system that integrates warehouse management, cargo management, warehouse output and input management, customer statistics, and warehouse output and input statistics. It manages cargo going in and out of the warehouses, customer relationships, warehousing equipment, warehousing information monitoring, inventory stocktaking, warehousing finance, basic business data, business data analysis, etc.

3) Intensive distribution system

This mainly analyzes and processes information in order processing, stocking, storage, picking, distribution, and delivery according to the needs of future distribution business. It comprises of basic information management, distribution plans management, cargo sorting management, inventory management, order management, and statistics and analysis management subsystems.

4) Freight information service and release system

This is an integrated and smart logistics information system oriented at the entire logistics system, including the management and release of traffic information, freight information, vehicle information, and warehousing information as well as cargo tracking information query, freight transaction matching services and user credit management.

5) Logistics financial service system

This performs digital management of financial services such as fund settlement, payment, credit, and insurance to help financing companies solve high capital turnovers, increase income sources, improve risk control capabilities and capital turnover rates, and to effectively organize and adjust capital activities. This ensures the smooth development of financial leasing, warehouse receipt pledges, bill pledges and bonded warehouse financial services for common interest of multiple parties.

6) Logistics value-added service system

This has six subsystems of information management: vehicle spare parts, vehicle maintenance, vehicle insurance, fuel power, vehicle leasing, and additional support. They cover vehicle repair, maintenance, insurance, power and the whole process of personnel work activities, thus guaranteeing the safe operation of logistics and perfecting the smart platform.

7) Safety management and emergency support system

In view of the data and system security of the smart platform, a safety management and emergency support system is established to improve business management. As soon as an emergency strikes, contingency will be initiated at the fastest response to minimize losses and guarantee corporate information and business security. It mainly includes a business security management system, an information and system security management system, a security evaluation system, a security early warning system, and a contingency plan system.

8) Big data application service system

Supported by technologies such as the IoT, cloud computing, data warehouse, data mining, geographic information system, and business intelligence, it effectively summarizes, analyzes and processes the data of a company's daily business operations, displays the results in the form of charts, and prepares reports automatically for decision-making. It improves the management level and operational efficiency. At the same time, it releases data to related customers and entities with business dealings to cultivate customer loyalty and assist them in making decisions.

2. Bulk commodity supply chain information service platform

This enables information sharing between the supply chain node enterprises, weakens the bullwhip effect, unifies their decision-making and refines their management, thereby improving the entire supply chain's operation and management. It has seven component systems: cooperative enterprise management, procurement service information, an inventory service information, sales service information, supply chain financial service, supply chain collaborative management information, and a big data application service.

1) Cooperative enterprise management system

Based on demand for customer management in the construction of the bulk commodity supply chain information service platform, a cooperative enterprise management system is established to perform classified customer management and improve efficiency. This includes six subsystems: basic information management, manufacturer management, supplier management, seller management, logistics service provider management and statistical analysis.

2) Procurement Service Information system

The procurement service information system comprehensively harnesses purchase application, purchase orders, incoming inspection, warehouse receipt, purchase return, purchase invoice processing, supplier management, price and supply information management, order management, and quality inspection management. This completes the two-way control and tracking of the entire procurement logistics and fund flow. There are six subsystems: basic information management, supplier management, procurement plan management, order management, receipt management and statistical analysis.

3) Inventory service information system

This adopts modern information, IoT, and big data mining technologies to integrate the warehousing and inventory management requirements of all supply chain node companies. It systematically collects, processes, transmits, stores, and exchanges inventory data for efficient delivery and sharing throughout the supply chain. It embraces five subsystems: basic information management, inventory management, order management, information release, and decision support.

4) Sales service information system

This integrates customer files, sales leads, sales activities, business reports, and sales performance summary. It consists of 5 subsystems: basic information management, customer management, order management, after-sales management, and decision support; the collaboration between them can optimize sales services.

5) Supply chain financial service system

This can render different levels of information services and auxiliary decision-making for different platform users to meet their demand for supply chain financial services. It collects, processes, organizes, stores, releases and shares supply chain financial information to integrate the overall chain, so as to reduce costs but improve efficiency. It comprises of five subsystems: basic information query, business operation, oil product finance, risk control, and decision support.

6) Supply chain collaborative management information system

It offers information-based management tools for the supply chain, assists in the information sharing mechanism between its upstream and downstream companies, and the coordinated operation of forecasting, planning, inventory, procurement, and sales services. It improves supply chain competitiveness while maximizing the benefits of the chain nodes. It contains five subsystems: basic data management, demand management, business management, interface management, and data sharing.

7) Big data application service system

As required by a company's supply chain management business, this combines data mining, IoT, and cloud computing to extract, collect, integrate, and share business data for each system to support and guarantee its operation on the business platform. It is made up of basic data management, predictive analysis, statistical analysis, operation analysis, and business intelligence subsystems.

3. Bulk commodity electronic trading platform

The bulk commodity electronic trading platform has 7 subsystems: bulk commodity category management; bulk commodity e-commerce; member service management; market information; trade financial; corporate credit and risk management, and big data application service. Together, they serve various businesses related to commodity trading. This platform enables interaction between enterprise and suppliers, sellers and customers as well online negotiations between business owners and customers, and online transactions and payments. Therefore, it achieves high-efficiency, low-cost information management, and networked operations in logistics and fund flow.

1) Bulk commodity category management system

It has five subsystems: basic information collection, category definition and identification, category catalog system construction, category management, and statistical analysis.

2) Bulk commodity e-commerce system

Its ten subsystems are commodity display, e-transaction, contract management, order management, settlement management, transaction matching, transaction information management, user management, customer service, and statistical analysis.

3) Member service management system

There are five subsystems: member information management, member business service, member information service, system configuration management, and statistical analysis.

4) Market information system

It is made up of five subsystems: market center, market information release, market analysis, market consulting, and statistical analysis.

5) Trade financial system

It is composed of 7 subsystems: basic financial information management, trade financing, commercial factoring, loan management, credit assurance, financial management, and statistical analysis.

6) Corporate credit and risk management system

It contains five subsystems: corporate basic information management, standardized credit reporting system management, corporate credit reporting management, credit rating management, and risk management.

7) Big data application service system

It consists of six subsystems: data warehouse management, data development, statistical analysis, predictive analysis, operational analysis, and intelligent business.

4. Logistics park information service platform

The logistics park information service platform is divided into seven major systems: basic information service, enterprise information service, enterprise business, industrial service, e-government, public service, and the big data application service. The key function of the platform is to integrate, manage, and share information resources in the park. It aims to create an advanced technology demonstration platform and informatized management platform to serve logistics gathering, resource sharing and long-term development.

1) Basic information service system

It refers to the management provision of park layout, function, personnel, credit, park organization, and facilities for park customers in accordance with the philosophy of openness and resource sharing.

2) Enterprise information service system

It refers to the provision of information inquiry, technical innovation, quality inspection, management consulting, entrepreneurial counseling, market development, personnel training, financing guarantee, environmental governance, and modern logistics for park customers.

3) Enterprise business system

It is a set of simple and practical office automation solutions based on fixed and mobile networks to help the park and customers efficiently use the Internet and wireless networks to build an integrated collaborative office platform and convenient and fast communication channels. This enables efficient work within the park and between them, meets the needs of key customers for the reporting of significant data, information and major emergencies, and creates a collaborative business circle.

4) Industrial service system

An aggregation is formed based on the distinct types of logistics in the park whose development is better managed by building a logistics service system. Targeted at them, corresponding services as well as management and technical function systems are set up to serve the development of regional logistics.

5) E-government system

It uses the information platform of the logistics park to integrate information resources of the park and its customers, building mutual information exchange and work management channels.

It thereby forms an overall information sharing advantage and work management mechanism, strengthening communication, and improving work efficiency and informatized management.

6) Public service system

It is an important carrier and way to realize public service of the park. It assumes the responsibility of collaboration and communication for customers, skilled workers and park managers, and plays an important role in promoting logistics development and improving the development environment. The system promotes the optimized configuration of resources of the park and specialized division of labor and collaboration, thus benefiting the development, transfer and application of key common technologies. It has various functional modules: financial support, public talents, technical innovation, e-commerce, facilities and equipment, and public business service.

7) Big data application service system

Based on the core business data, basic business data and public business data of the park business system, by adopting modern data methods and technology, it conducts detailed analysis and mining of the logistics park information data. Specifically, it performs data mining and immediate query, multi-dimensional analysis, and decision support. It can provide comprehensive and multi-level decision support and knowledge services through automated indexes and report information management.

5.3.5 Intelligent decision support layer

Application technology of IoT is adopted to combine the collected data information with the database technology and mining tools to assist decision-makers in predictive analysis, statistical analysis, operational analysis, and business intelligence. The environment of decision-making processes and plans is simulated, allocating various information resources and analysis tools to help decision-makers improve a company's intelligent decision-making and service quality, and help enterprise and relevant entities in the logistics park realize intelligent management.

5.4 Design of construction plan

Combined with the construction background and positioning of the information platform and based on the informatization foundation and demand analysis of logistics, the overall construction plan of the smart platform is designed and guided by the ideas of the information platform. It aims to comprehensively and systematically guide informatization progress, coordinate the application of information technology, meet the development needs of logistics in a timely manner, make full use of resources, and improve investment effectiveness of informatization, the level of future

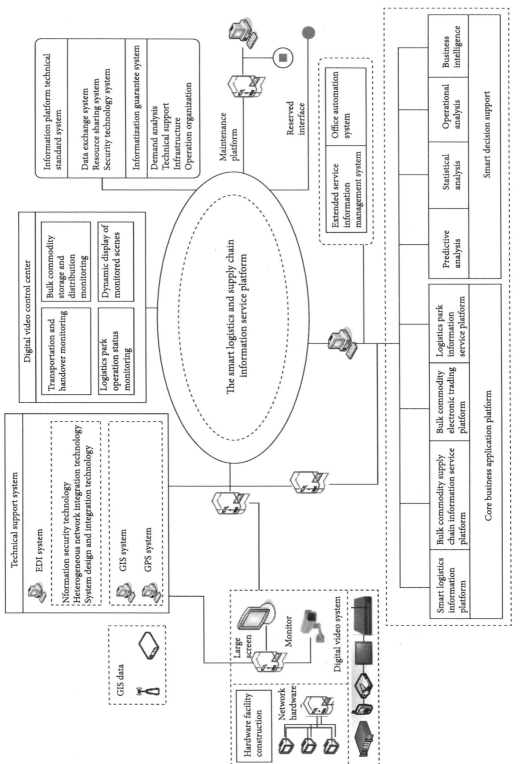

Figure 5-3 The overall plan for the construction of the smart logistics and supply chain information service platform

production control and management, operation and decision-making. It also gives full play to advantages of modern logistics clusters and enhances service quality of enterprises' integrated supply chain. This is shown in Figure 5-3.

The construction plan, combined with the development needs and strategies of the four major business clusters of enterprise, is carried out from the four platform levels of the basic support environment, technical support and guarantee system, core business application system, and standard specification system. The core business application platform embraces four information platforms: smart logistics information, bulk commodity supply chain information, bulk commodity electronic trading, and logistics park information service.

5.4.1 Main construction content

The construction plan of the smart platform, based on the informatized management needs of integrated logistics, bulk commodity supply chain service, bulk commodity trading, and logistics park service business clusters, combines the requirements for refined process and business management. The primary construction tasks are computer communication network design, smart logistics and supply chain business management platform design, real-time dynamic digital monitoring center design, system integrated solution design, and information platform technology and planning system design. Details of the construction are enumerated as follows.

1. Computer communication network design
Design the computer network structure and equipment, data transmission and communication interface, data link control, public data exchange network, and the Internet structure; integrate the scattered resources on the Internet into an organic whole, and fully share resources and collaborate so that relevant personnel can transparently make use of resources and obtain information on demand.

2. Smart logistics and supply chain business management platform design
Combine the core and auxiliary businesses of the four major industrial clusters to design the application information system according to its attributes and needs. That is, design the positioning, layout, and functions of each application system in the platform. The smart platform embraces the bulk commodity supply chain information service platform, the bulk commodity electronic trading platform, and the logistics park information service platform.

3. Real-time dynamic digital monitoring center design
Combine the requirements of integrated logistics business for refined and digital production to design a digital monitoring system with real-time dynamic monitoring, display and information release as the core, and conduct centralized management of scattered and self-contained monitoring systems. This is characterized by being real-time distributive and synchronized. It enables technical managers to use real-time monitoring and business analysis for the operating environment of

transportation handover, bulk commodity storage and distribution, and logistics park operation, respond quickly to problems that arise in the business process, and assist management and decision-making.

4. System integrated solution design

Comprehensive integration technology mainly embraces system data integration, system environment support, business management and decision-making, standardization, enterprise modeling, and system development and implementation. As great support for the smooth completion of comprehensive integration, it is more than just simply linking various technologies together, but organically combining the separated subsystems so that they can work in coordination for optimal performance.

5. Information platform technology and planning system design

Through demand analysis of the information platform and design of the business application system, the core and auxiliary business processes are analyzed, the data and information flow of each process are integrated, and data exchange, resource sharing, and security technology information platforms come into shape.

5.4.2 The overall framework of the construction plan

Combined with the need of the four major industrial clusters for business informatization, the overall framework of the smart platform is built according to the general construction plan of an information platform to comprehensively and systematically guide informatization and coordinate and develop the application of information technology. This can fully meet the needs of users and functions and reasonably configure equipment, information, and human resources of a company. The overall framework to construct the smart platform is shown in Figure 5-4.

The plan is implemented from six levels to propel informatization. The details are presented as follows:

1. Basic support environment

This level constructs a system software and hardware environment related to the information platform, the promotion of company informatization, and basic support for digital video surveillance systems. Regarding the hardware environment, there is the construction of the master computer, large screen, multimedia system, automated office system network, and server and surveillance contact base station; the software environment renders model selections of the service agreement framework and operating systems for core business systems and digital video surveillance systems.

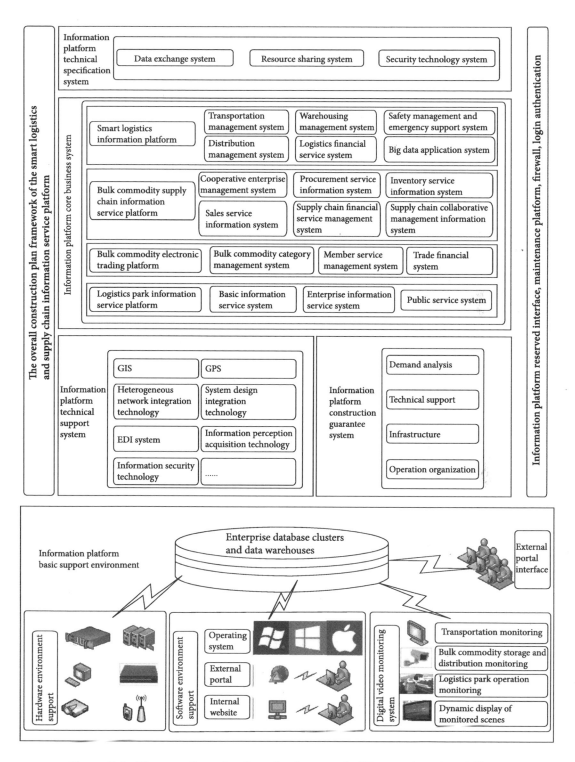

Figure 5-4 The overall construction plan framework of the smart service platform

2. Digital video surveillance system

Generally put into use upon the initial completion of the information hardware infrastructure and the engagement in the bulk commodity supply chain business. The construction of a digital video surveillance system for real-time monitoring can enable the four major industrial clusters to control and safely run the entire business process, and perform digital and dynamic full-scene monitoring of key operations such as bulk commodity storage and distribution and logistics park operations.

3. Technical support system

This is built on the basis of ready platform hardware facilities and equipment, a complete software environment, and gradual improvement of computer and communication networks. It mainly includes information acquisition technology applied to loading, unloading and distribution, and GPS, GIS (geographic information system), system development and integration, heterogeneous network integration technology, and EDI (electronic data interchange) system and information security technology applied to office automation system.

4. Guarantee system

The establishment of a guarantee system for the platform improves progress of the links such as the preliminary preparation, infrastructure construction, system design and comprehensive integration, and platform application implementation. The information platform's role in enhancing informatization of the four major cluster businesses is therefore vividly depicted. And the guarantee system mainly includes demand analysis, technical support, and infrastructure and operational organization.

5. Core business system

Combined with the development needs of bulk commodity logistics and supply chain informatization, the smart platform is designed per the development concept of the four core industry cluster platforms: smart logistics information; bulk commodity supply chain information service; bulk commodity electronic trading, and the logistics park information service.

6. Technical specification system

To further coordinate a company's various businesses, adjust application specifications of the information platform technology, and improve overall informatization, a technical specification system is established. This is based on the construction of the four core sub-platform business application systems. It includes data exchange, resource sharing, and a security technology system.

5.4.3 Features

The smart service platform is constructed with highly unified planning and design. The overall structure enables the interconnection of information – data is highly shared and has sufficient flexibility. It can both integrate the four major business information platforms and grant access to newly developed application systems.

1. Great integration

The smart platform integrates separate equipment, functions, and four major business information platforms into an interconnected, unified, and coordinated platform through a structured comprehensive management system and computer network technology for resource sharing and efficient management.

2. High-performance communication network

The smart platform adopts an integrated mobile network solution, both wired and wireless, which fully assesses the existing network, thus meeting the need for real-time tracking, improving platform management and control capabilities, and integrating network hardware and software design management.

3. Application of new technology

The construction of the smart platform is supported by advanced technologies such as IoT, cloud computing, mobile IoT, and big data mining. It cooperates with suppliers to share software and hardware resources and information to reduce hardware infrastructure investment while improving business data processing capabilities and dynamic resource expansion capabilities.

CHAPTER 6

SMART LOGISTICS INFORMATION PLATFORM

6.1 Overview

6.1.1 Significance and goals

As the logistics industry transforms and upgrades, the demand for smart logistics has become stronger and more diversified. This includes demand for services in logistics data and cloud technology. In July 2016, the Ministry of Commerce issued the Notice on Determining Smart Logistics and Distribution Sample Flats to demonstrate the construction of such a system. Smart logistics is the cornerstone of the *Made in China 2025* strategy. According to its current rapid progress, it is expected that by 2025 the market size of smart logistics services will have exceeded one trillion RMB.

The goal of building a smart logistics information platform is to adopt smart technologies (such as big data, cloud computing, and smart hardware) as support to enable system perception in all aspects of logistics transportation – warehousing, packaging, loading and unloading, circulation processing, distribution, and information services.

In the meantime, building this platform is conducive for thinking, perceiving, learning, analyzing and deciding, and intelligently executing in logistics. It also enhances its intelligence and automation level; it benefits the integration of supply chain logistics system resources, discovers, summarize, and innovates. Thus, it realizes logistics regulation intelligence, discovery, and innovation; it reduces social costs and raises efficiency, thus promoting the development of high-end products, and the transformation and upgrading of the logistics industry.

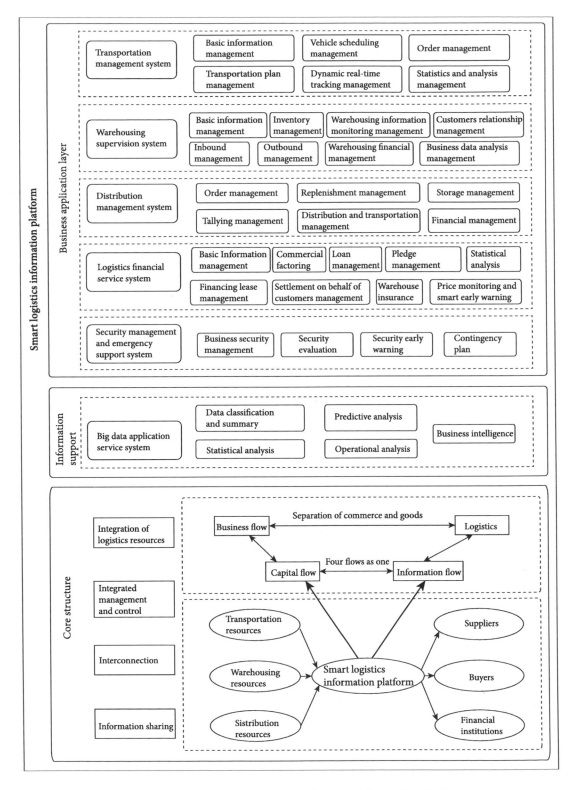

Figure 6-1 Framework of the smart logistics information platform

6.1.2 Overall Framework

Smart logistics has two key features. One is interconnection and data-driven, which requires logistics elements to be interconnected and digitized, and data to drive all insights, decisions, and actions; the other is in-depth collaboration and efficient execution – that is, cross-group and cross-enterprise, and cross-organization in-depth collaboration based on the smart algorithm of global optimization, to schedule the efficient division of labor and collaboration of all participants.

The smart platform is an effective tool that reflects the advantages of smart logistics and implement the informatization of enterprise and industry. Combined with development strategy, the needs of informatization, the basic requirements of logistics, and the core business of logistics companies, it embodies the integration, informatization, and collaboration within smart logistics. It is divided into eight subsystems: transportation management; warehousing supervision; distribution management; freight information service and release; logistics financial services, logistics value-added service; security management and emergency support, and big data application service.

This platform enables interconnection and information sharing between various application systems as well as industrial collaboration within the service area to reduce costs, improve business efficiency and management, and render centralized, integrated, standardized, and intelligent information services for business management and control. The framework is shown in Figure 6-1.

6.2 Transportation management system

6.2.1 Construction requirements and significance

1. Construction requirements
Transportation is an major part of logistics operations as transportation time and costs account for a considerable proportion of each link. Modern transportation management is the management of transportation networks and operations. It conveys information regarding transportation tasks, resource control, status tracking, and feedback of different regions within the networks. It is able to adopt modern computer technology and logistics management methods to smarten and visualize transportation management, so as to diminish low efficiency, poor accuracy, high costs, slow response, and inability to meet customer needs.

Specifically, it uses modern information technology to effectively manage transportation planning, means, personnel, process tracking, dispatch and command, and to solve problems in smart integrated transportation. Efficiency, convenience, and individualized needs are taken as the yard sticks; the complementary promoting effects of various transportation methods are integrated to enable efficient operation of the entire transportation system; simultaneously, it coordinates the relationship between various transportation methods to further improve capacity, speed, and economic benefits.

2. Significance

The transportation management system, based on the status quo of logistics and combined with the development goals of the companies, has long-term goals. Starting from the macro-level of improving transportation management information, it adopts new technologies, new methods, and new equipment to share and coordinate information resources between the business departments of logistics for unified management of transportation. It renders first-class management, organization and service.

It mainly manages transportation tools and processes, which is conducive to improving service. Regarding smart management of transportation, it effectively reduces management costs, improves service quality during transportation, guarantees the safety of vehicles and goods, and gives certain references for decision-making support; in terms of efficient operation, it controls vehicles, personnel, and the completion of transportation tasks in real time, allocates task resources reasonably, avoids idle vehicles, personnel and fleets when there are intensive transportation tasks, efficiently completes transportation tasks, and increases effective vehicle mileage; when it comes to the integration of social vehicle capacity, it improves management efficiency of fleets and vehicles, reduces management costs, and enhances service quality with the help of information technology and smart management methods.

6.2.2 Business process analysis

This takes the completion of transportation tasks as the core goal by deploying waybill information, vehicles, personnel, and goods; receiving requirements are clarified, transportation plans are drawn, vehicles are scheduled, transportation is tracked and core links are fedback. Therefore, this system should, on the basis of meeting customer needs, perform informatized and smart management in the transportation process to improve service quality, increase system flexibility and adaptability, and reduce transportation costs. It should also coordinate transportation plans, including plan formulation, plan transmission, feedback modification, etc.; in the transportation link, processed operation and management of cargo handling, in-transit transportation, corresponding document reception, and information transmission should be performed to achieve refined control. Through real-time tracking of relevant information and timely online updates, information on vehicles and goods in transit is fed back in order to supervise the transportation business. This is depicted in Figure 6-2.

In detail, the transportation business process goes as follows: the business department prepares the transportation plan and delivers the feedback adjusted plan to the dispatch center; this deploys vehicles and personnel according to the plan and generates the corresponding documents; the transportation information, with the documents as the carrier, flows between departments; the real-time information of goods and vehicles is fed back through the positioning function of the corresponding technology, thus transportation is completed efficiently, quickly and safely.

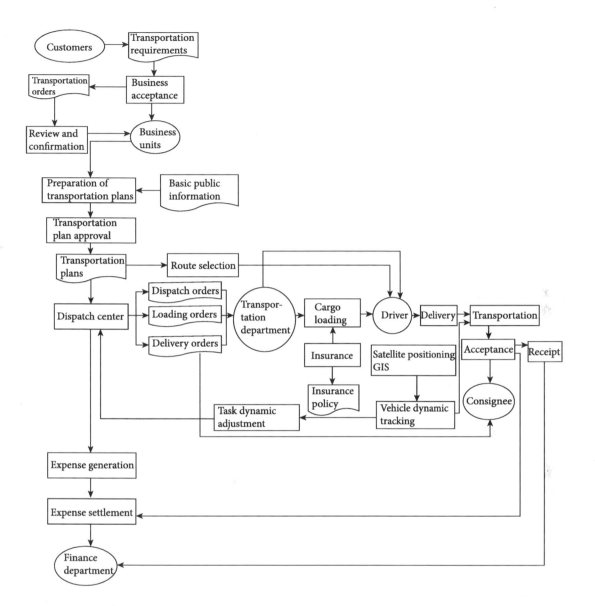

Figure 6-2 The transportation business process of logistics companies

The transportation business data process builds a physical model of the internal data flow. It is a mode of data transmission between departments to reflect the actual transportation business data processing. Combined with the analysis and sorting of the actual operation requirements during transportation, the data process of core business links is taken as the core data process of the transportation management system. This is shown in Figure 6-3.

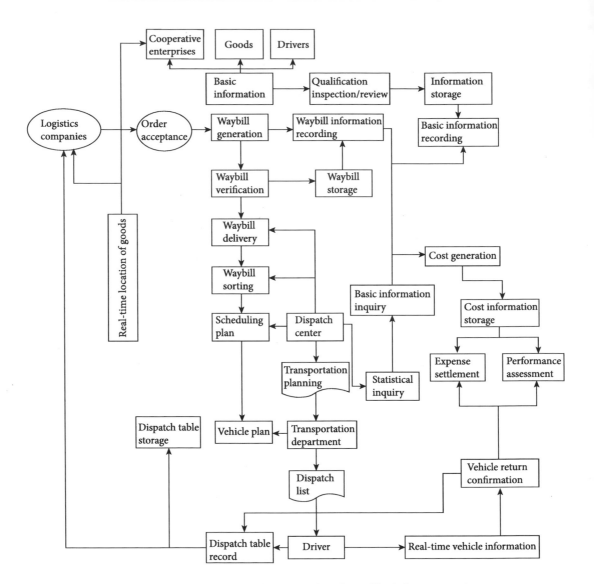

Figure 6-3   Transportation business data flow of logistics companies

6.2.3 Overall structure and function description

The transportation management system covers core business related to logistics transport. It is an important link and entry point for improving the comprehensive capabilities of a company, reducing transportation costs, and exploring economic growth. Its functional architecture embraces subsystem management for basic information, transportation planning, vehicle dispatch, dynamic real-time tracking, vehicle status and safety, finance and performance, and statistics and analysis, as shown in Figure 6-4.

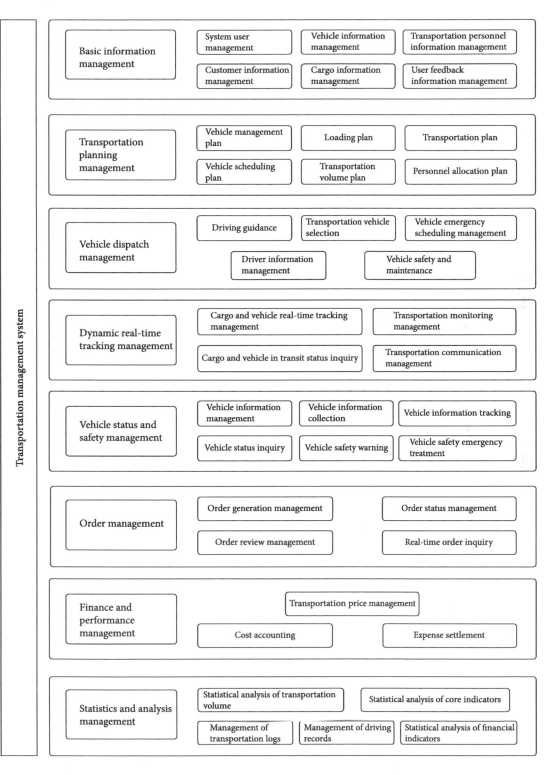

Figure 6-4 The functional architecture of the transportation management system

1. Basic information management subsystem

It has the following functional modules of management: system user, vehicle information, cargo information, transportation personnel information, customer information, and user feedback information. It aims to organize and manage information regarding business, vehicles, and user feedback, and offer support and convenience in planning and transportation as well as the circulation and transmission of information within departments.

2. Transportation planning management subsystem

It has the following functional module plans: vehicle management, loading, transportation, vehicle scheduling, transportation volume, and personnel distribution. The goal of this subsystem is to integrate, classify, and redistribute transportation requirements to carry out preliminary planning with vehicles, personnel and business, and formulate corresponding plans to guide vehicle scheduling. This greatly improves efficiency, better guides vehicle scheduling, and ensures the smooth completion of transportation tasks.

3. Vehicle dispatch management subsystem

It has the following functional modules: driving guidance, transportation vehicle selection, emergency vehicle scheduling management, driver information management, and vehicle safety and maintenance. According to transportation tasks and plans, dispatch management is performed for self-owned vehicles and public vehicles to form an organic whole to maximize transportation potential; meanwhile, according to cargo flow, flow direction, and seasonal changes – as required by the transportation plan – vehicle transportation tasks are carefully arranged to ensure a safe, efficient and fast completion.

4. Dynamic real-time tracking management subsystem

It has the following functional modules: real-time cargo and vehicle tracking management, transportation monitoring management, cargo and vehicle in-transit status query, and transportation communication management. The purpose of this susbsystem is to manage in-transit vehicles and their information through dynamic real-time tracking and feedback from vehicle and cargo transportation status. Additionally, it is to manage and control in-transit vehicles, on-board terminals, transport personnel, transportation cargo location, delivery time, and cargo status for seamless connection with vehicle scheduling; thus, transportation information transmission forms a fully closed loop.

5. Vehicle status and safety management subsystem

It has six functional modules: vehicle information collection and management, vehicle information tracking, vehicle status query, vehicle safety warning, and vehicle safety emergency processing. It is designed to manage the vehicle status and feedback information regarding whether the vehicle is under maintenance, whether its due maintenance and repair, its assessment results, and safety warnings when vehicles are unqualified.

6. Order management subsystem

It has the following functional modules: order generation management, order status management, order review management, and order confirmation query. It aims to informatize the entire process of the actual transportation business, including order generation, development, establishment, confirmation, completion, and information storage, while maintaining the integrity of the transportation business data process.

7. Finance and performance management subsystem

This calculates transportation costs, evaluates performance and allocates personnel and drivers. This enables control of transportation prices, assistance in performance management, generates incentives, and effectively manages the company's business. The leadership allows better understanding of business affairs and making business decisions.

8. Statistics and analysis management subsystem

It has five functional modules: transportation volume statistical analysis, transportation log management, driving record management, financial indicator statistical analysis, and core indicator statistical analysis. It aims to read, classify, analyze, and compute various data generated every day to assist a company's decision-making and provide advice for companies that render logistics services.

6.3 Warehouse supervision system

6.3.1 Construction requirements and significance

1. Construction requirements

This integrates inventory management, cargo inbound and outbound management, customer statistics, and makes full use of big data and smart tech such as data warehousing, sharing, and mining. It improves warehousing operation efficiency, reduces operating costs, achieves transparency of business processes, and ensures efficient processing, effective use, and timely information sharing. Also, it can employ smart terminals and information platforms to perform real-time statistics and data analysis on warehousing operations to perform instant operational adjustment.

2. Significance

Logistics companies use modern information technology to build a corresponding warehousing supervision system to collect, classify, transmit, summarize, identify, track, and search information generated in the logistics process. Data between various information systems is transferred for sharing, and applied throughout the business processes of warehousing activities. Thus, all-round control is implemented in goods flow and the storage processes to improve efficiency, increase processing speeds, standardize businesses, reduce storage costs and cargo damage, enhance service quality and informatization, and ensure the efficient circulation and interconnection of information. Consequently, logistics and warehousing business is intelligentized, informatized, and automated and thus has stronger core competitiveness.

6.3.2 Business process analysis

The warehousing supervision business process revolves around the storage and management of warehouse goods. It is a series of operations – from the arrival of goods (to the warehouse) until they are shipped out. Warehousing, as an important link in the entire supply chain, bridges manufacturers and consumers. And the business process of warehousing plays a decisive role in improving efficiency and optimizing resource allocation. Therefore, the system should, on the premise of meeting customer needs, divide the core business processes to improve service quality, increase system flexibility and adaptability, and generate more value.

After analyzing three core businesses of warehousing supervision: cargo inbound management; internal inventory management; and cargo outbound management, it can be learned that cargo inbound management includes cargo inspection, inbound operations, inbound inquiry, and allocation of holds; inventory internal management mainly involves goods inquiry, goods inventory, holds adjustment, holds information management; while cargo outbound management performs goods inspection, outbound preparation, picking and stocking, and outbound order generation. These processes are shown in Figure 6-5.

6.3.3 Overall structure and function description

As the hub of logistics information, the warehousing supervision system is a key link of controlling inventory, reducing costs, and improving economic benefits. To ensure its smooth development, it should carry out effective inventory management based on the needs of upstream and downstream companies, engage in effficient inbound and outbound operations, and offers information support to supply chain node enterprises for decision-making. The functional architecture of the warehousing supervision system embraces basic information, inbound, inventory, outbound, warehousing information monitoring, financial, customer relationship, and business data analysis management, as shown in Figure 6-6.

1. Basic information management subsystem
It has four functional management modules: access setting, user information, inventory information, and cargo information. It is mostly applied to manage and tally basic information in the warehouse supervision system as well as to provides access management to various systems. The information in this subsystem runs through the entire warehousing supervision system, and is the basis for specific businesses such as cargo inbound management, internal inventory management, outbound management, and data analysis.

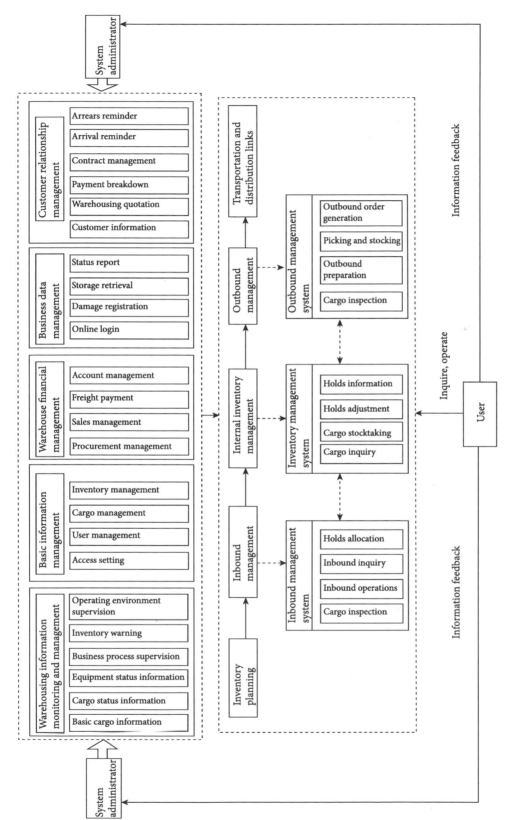

Figure 6-5 The business processes of logistics companies' warehousing supervision

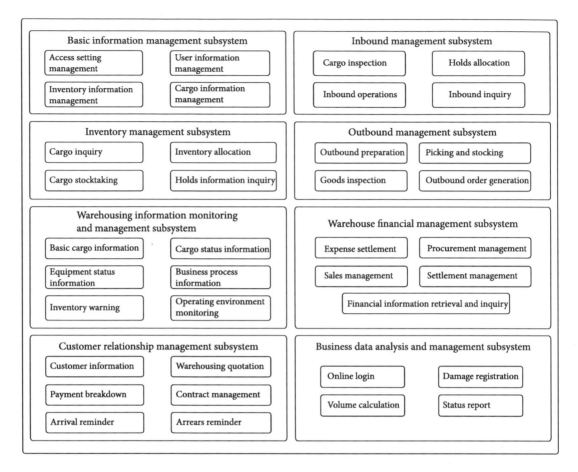

Figure 6-6   The functional architecture of the warehouse supervision system

2. Inbound management subsystem

It has four functional modules: cargo inspection, holds allocation, inbound operations, and inbound inquiry. It is used mostly to manage and record preparatory work and inbound operations of goods. The accuracy of the goods is checked first according to the purchase order and then the system allocates the corresponding storage holds, time and operators. This standardizes inbound operations.

3. Inventory management subsystem

It has four functional modules: cargo inquiry, inventory allocation, cargo stocktaking, and hold information inquiry. It is often employed for the management and inquiry of goods in the warehouse. Therefore, users are provided with the latest cargo inventory information and goods between multiple warehouses are allocated to adapt to regulatory requirements of multiple good types and warehouse environments.

4. Outbound management subsystem

It has four functional modules: outbound preparation, picking and stocking, goods inspection, and outbound order generation. It is generally adopted to manage and record preparation work and outbound operations before goods are shipped out of the warehouse. According to the names and quantity of goods that customers require, they are inspected by staff to determine whether inventories are sufficient. The system will arrange the picking, preparation, and outbounding of the goods according to the shipment time and type. This standardizes the process.

5. Warehouse information monitoring and management subsystem

It has six functional modules: basic cargo information, cargo status information, equipment status information, business process information, inventory warning, and operating environment monitoring. It is mainly utilized for preliminary processing, display and inquiry of various warehousing information as well as classification and real-time data update in other subsystems. This comprehensively monitors the goods in storage, in transit, and in transfer, and improves the quality of warehousing services.

6. Warehousing financial management subsystem

It has five functional modules: expense settlement, procurement management, sales management, settlement management, and financial information retrieval and inquiry. In accordance with relevant laws and regulations, it runs statistics and calculations on the storage costs, rental expenses, and the purchase and sales needs of the companies. It also organizes their financial activities in accordance with financial management principles, handles financial relations, analyzes settlements, and prepares relevant financial statements and business operation diagrams. This is done to automate financial affairs management of warehousing companies and support decision-making.

7. Customer relationship management subsystem

It has six functional modules: customer information, warehousing quotation, payment breakdown, contract management, arrival reminder, and arrears reminder. It conducts statistical analysis and customer evaluation of historical business data through the customer relationship database to analyze customer needs, render personalized services, improve service quality, and support decision-making.

8. Business data analysis and management subsystem

It has four functional modules: online login, volume calculation, damage registration, and status report. Combined with big data and intelligent technology, it systematically analyzes all data in warehousing supervision and feeds back multiple information data to system operators in the form of reports and charts, with a focus on improving the service quality and increasing warehouse utilization.

6.4 Distribution management system

6.4.1 Construction requirements and significance

1. Construction requirements

Distribution is directly oriented at consumers, thus most directly reflects the service quality of the supply chain. How to provide products and services at the right time and place with higher service and quality at lower costs has become a distribution problem that logistics companies must consider. Multi-category, low-volume, and multi-frequency distribution makes higher requirements on a company in service quality, resources and cost. And through a variety of information technology methods, the intensive distribution system revolves around integrated management of distribution. Ultimately, it reduces costs but increases delivery efficiency.

2. Significance

The adoption of smart distribution leads to a centralized and modernized production organization form. It is able to coordinate the current transportation resources of logistics companies, give full play to the advantages of scale, explore alternative profit sources, and reduce logistics costs. It also makes full use of information, network technology, and modern organization and management methods, thus extending the scope of services in supply chain management, and integrating various resources to minimise waste.

6.4.2 Business process analysis

Distribution starts from the order entrusted by the customer, including order processing, purchase, storage, sorting, circulation processing, assembly and shipment, delivery, etc. This management process is shown in Figure 6-7.

As shown in Figure 6-7, the major business process issues distribution requirements for customers and dispatches vehicles once the transaction is accepted, while setting prices for goods and transferring information to the warehouse. If out of stock, replenishment takes place; sorting personnel and goods in the warehouse according to orders; also the delivery and transportation department arranges vehicles, selects routes, and monitors vehicles and goods in transit; once goods arrive at their destination, the consignee checks it before signing for confirmation.

6.4.3 Overall structure and function description

The distribution management system analyzes and processes information in order processing, stocking, storage, picking, distribution, and delivery. It consists of six management subsystems: order, replenishment, storage, tallying, distribution and transportation, and finance. The functional architecture is shown in Figure 6-8.

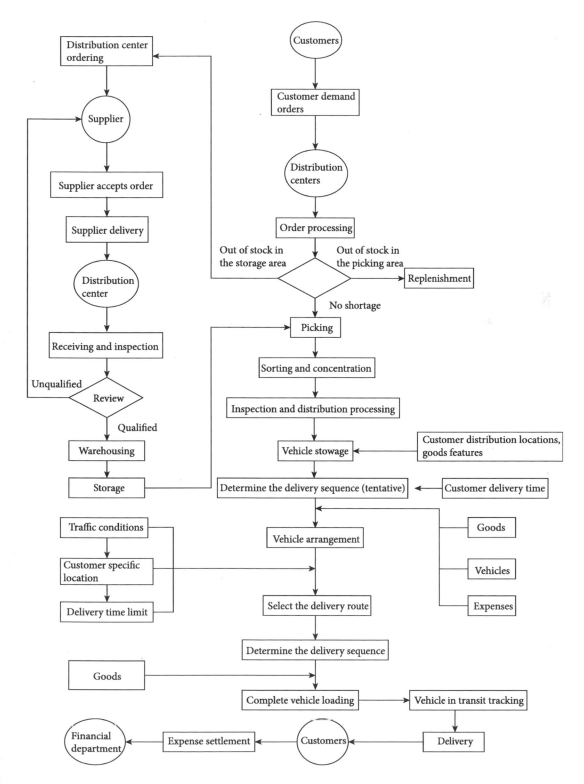

Figure 6-7 The process of a logistics company's distribution management system

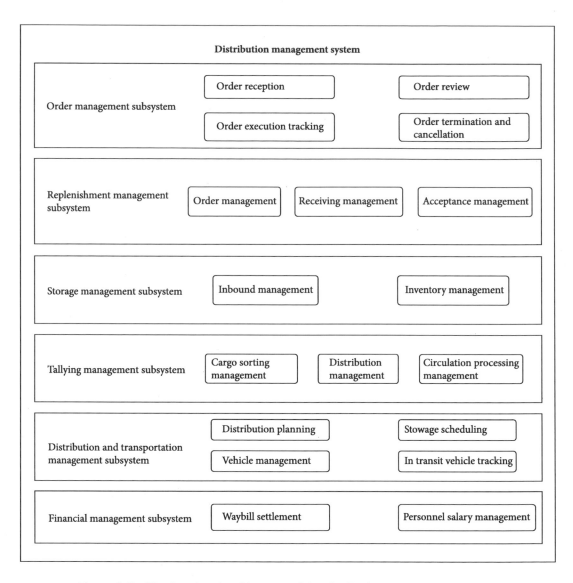

Figure 6-8   The functional architecture of the distribution management system

1. Order management subsystem

This includes customer order reception, review, execution tracking, and termination and cancellation to ensure customers are provided with thoughtful service. Order reception is about receiving customer orders and registering order information. According to order information, customer distribution, product nature, number of types, and delivery frequency are analyzed to determine the type, specification, quantity and delivery time of goods, record information, and draw replenishment plans in time. Order review examines order effectiveness – if it does not meet the standard, it will be modified or rejected. Order execution tracking follows orders and monitors their processing status.

2. Replenishment management subsystem

Replenishment management prepares for distribution, which involves raising supply, ordering and related quality inspection, handover, etc. It embraces order, reception, and acceptance management. The distribution center, according to good type ordered, replenish it from suppliers in a timely manner. It is also possible to place orders with suppliers in advance based on customer demand forecasts. Ordering the appropriate quantity not only meets customer needs, but also avoids inventory backlog as much as possible. Then the delivery time, location, product type, and quantity of different suppliers are tracked and managed; manpower and material resources are arranged in advance to receive the goods. Finally, as per the contract terms and related quality standards, type, specification, quantity, quality, and packaging are inspected and accepted. Upon acceptance, accouting, information input, and cargo inbound procedures are performed for goods to enter warehouses.

3. Storage management subsystem

It includes inbound and inventory management. Inbound management is scheduled and actual inbound data processing. Scheduled inbound data processing refers to scheduled inboud date, type and quantity of goods on the purchase order, supplier notification of arrival date, type and quantity of goods in advance, and regularly published schedules of inboud data reports. The actual inbound data processing records acceptance information of storage goods and how unexpected situations are handled according to purchase order number, manufacturer name etc. Inventory management includes goods classification, determiation of order quantity and time, inventory tracking management, stocktaking, and early warning management. The classification of goods counts the inventory by category and sorts and classifies them by quantity. Determination of order quantity and time refers to order quanity and time calculations based on goods names, unit price, existing inventory information, purchase lead time, and delivery cost. Inventory tracking management extracts information of the storage location, storage area, and distribution status of the existing inventory from the database, and generates inquiry reports for cargo, inquiry, inventory backlog, etc. Stocktaking management regularly records various goods stocktaking schedules, inputs stocktaking data, and prepares inventory profit and loss reports, etc. Early warning management is an alert system that tracks good numbers goods in stock, quality, and unsalable items.

4. Tallying management subsystem

It comprises of sorting, distribution, and circulation process management. Sorting management is based on customer order requirements and distribution plans, thus the distribution center quickly and accurately picks out goods from its storage point for classification and concentration. Reasonable planning and management of sorting leads to the improvement of operation efficiency and the reduction of operation costs. To make full use of the volume and load capacity of the transportation vehicle (to improve transportation efficiency), it is possible to have the goods of different users transported together by the same vehicle. Therefore, it is necessary to assemble them before shipment. Effective assembly can reduce delivery costs and traffic flow, thus alleviating traffic congestion. Circulation process management draws picking plans and circulation process plans based on customer orders, records the actual work of pickers or circulation processing personnel, and prepares and prints work reports.

5. Distribution and transportation management subsystem

This involves distribution planning, stowage scheduling, vehicle management, and in-transit vehicle tracking. Distribution planning clarifies distribution material categories, specifications, packaging, volume and delivery time, and formulates corresponding plans regarding delivery time, loading method, etc. Stowage scheduling refers to distribution scheduling tasks according to the actual situation of transportation resources, and to the generation of corresponding transportation instructions and assignments. Specifically, based on the weight of the goods, volume, destination, vehicle conditions, driver states, and route status, the optimal combination of vehicles, goods, and routes is decided. The stowage scheduling module has three functions: route selection, loading planning, and vehicle scheduling. Vehicle management includes performance statistics, file management, maintenance, consumption, route management, and maintenance management. In-transit vehicle tracking monitors vvehicle status on the road through GPS, and understands their location and status in real-time. It enables free inquiry about vehicle status in transit for any specified order.

6. Financial management subsystem

It consists of waybill settlement and personnel salary management. It settles completed waybills. Once goods are out of the warehouse, the distribution center prepares a receivable bill based on the shipment data and transfers it to the accounting department. As soon as the customer receives the goods, the order is complete and settled; the work status of personnel in each link of the distribution is collected, and the finance department has to issue a salary breakdown when paying them.

6.5 Logistics financial service system

6.5.1 Construction requirements and significance

1. Construction requirements

As the Chinese economy experiences adjusted development and state policies gradually loosen, logistics finance has become an important part of economic development – especially today when logistics is evolving rapidly. There is a huge market demand for logistics finance. Logistics companies should engage in finance such as warehouse receipt pledges, leasing, trade financing, customer settlement, commercial factoring, receivables and payables, vehicle trade repurchase, etc. They can effectively organize and allocate various types of deposits, loans, leases, insurance, discounts, mortage, precipitation funds, and the flow of funds for various logistics-related intermediate businesses handled by banks. This boosts the value of its own logistics business and create new profit growth points.

2. Significance

The financial service system can ensure the barrier-free circulation and sharing of logistics, information flow, and fund flow among logistics companies and financial institutions, thus improving information sharing; it can trace all financial service information from start to finish, including contracts, bills, loan issuance, fund flow, pledges, and insurance to ensure the interests of regulators, banks, manufacturers, and distributors. This improves the level of risk management and control; ultimately, it enhances efficiency, profitability, management of the companies themselves, and broadens financing channels; improves competitiveness of financial institutions, and the service level of the supply chain.

6.5.2 Business process analysis

Common forms of logistics financial services are financial leasing, warehouse receipt pledges, confirmation storage, etc. Their specific business process is analyzed as follows.

1. Financing lease

Financial leasing is an effective method of fundraising: It is a financial business in which the consumer purchases the leased property from the leaser and provides it for use. It has to go through applications for financial leasing, project evaluation, contracts signing, loan application, and contract execution.

2. Warehouse receipt pledge

This stores goods in the warehouse as required by the goods owner and applies for bank loans based on the warehouse receipt issued. The bank grants a certain percentage of the loan based on goods value; meanwhile, the warehouse supervises goods and collects a certain amount of remuneration on behalf of the bank. This process is shown in Figure 6-9.

3. Confirming storage

Confirming storage is a financial service that takes bank credit as the carrier and bank acceptance bills as the settlement tool when the bank controls the property in goods; the seller (or warehousing party) is entrusted to keep the goods and accept the bills; the amount other than the margin is guaranteed by the seller with the repurchase of the goods, and the banks provide acceptance bills to manufacturers (sellers) and their distributors (buyers). This is shown in Figure 6-10.

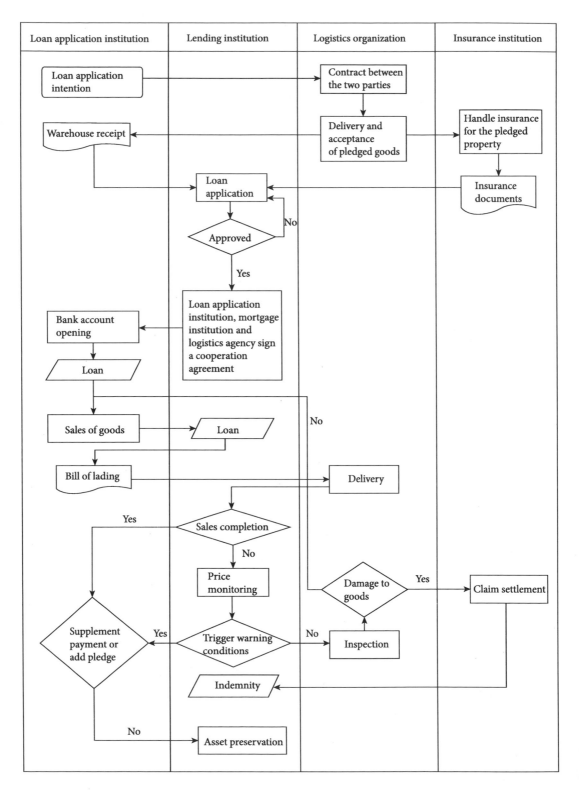

Figure 6-9 The business process of warehouse receipt pledge of a logistics company

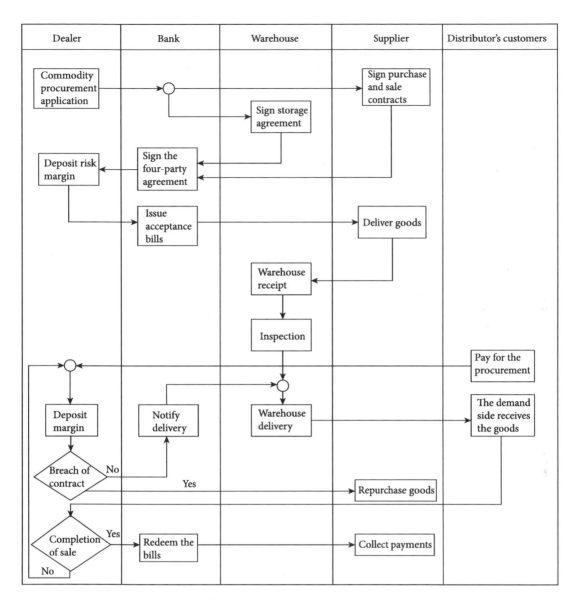

Figure 6-10 The business process of confirming storage of a logistics company

6.5.3 Overall structure and function description

As supervised by financial institutions, financial service systems, based on the needs of logistics financial business, design corresponding subsystems to manage various basic, bank loan, insurance, and pledge information; it then monitors the price of pledged items in a real-time manner. The management subsystems are basic information, financial lease, commercial factoring, customer settlement, loan management, warehousing insurance, pledge process, price monitoring, and smart early warning and statistical analysis. The functional architecture is shown in Figure 6-11.

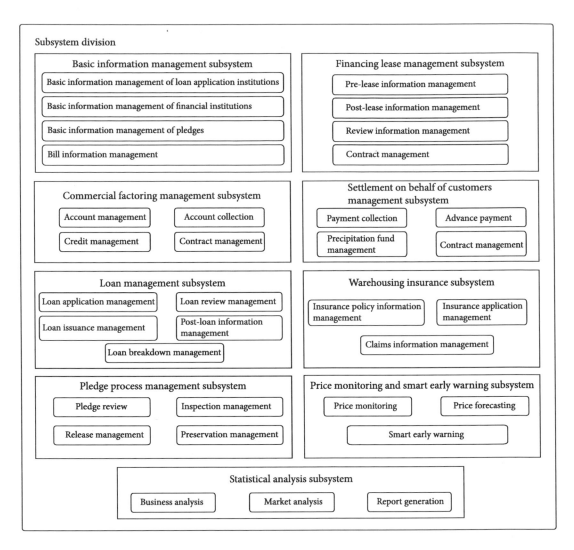

Figure 6-11   The functional architecture of the logistics financial service system

1. Basic information management subsystem

It manages the basic information of loan applicants, financial institutions, pledges, and bills. It inquires about, modifies, adds various information, and generates reports based on this, which lays the data foundation for business development and data analysis.

2. Financing lease management subsystem

It manages review information, pre-lease information, post-lease information, and contracts. It is normally used for vehicle post-sale leaseback business. It informatizes the management and control of each vehicle link post-sale leaseback and visualizes the management of leasing information and vehicle conditions, repayment information and property rights, guarantors, and suppliers, thus improving efficiency and management in financial leasing.

3. Commercial factoring management subsystem

It embraces account management, account collection, credit management, and contract management. It informatizes the management and control of these in commercial factoring, so as to reduce bad debts and improve business level and quality. It also achieves early risk management and control, clarifies critical time points, and tracks accounts throughout the process.

4. Settlement on behalf of customers management subsystem

It includes collection on delivery, advance payment, precipitation fund management, and contract management. It standardizes the management of collection on delivery and advance payment, and fully monitors the flow of precipitation funds. This provides customers with high-quality, efficient settlement services, thus reducing the risk of capital transactions and guaranteeing capital safety. It makes use of the turnover of precipitation funds to improve overall efficiency in the supply chain.

5. Loan management subsystem

It includes loan management of applications, reviews, issuance, post-loan information, and breakdowns. It manages loan application, review, issuance, and tracking inspection so that the management of each link and related business data is standardized and efficient. Consequently, the actual value and risk of each loan are accurately assessed, the flow of loans are tracked, loan risks are reduced, and credit increases.

6. Warehousing insurance subsystem

It includes insurance policy information management, insurance application management, and claims information management. It inquires and manages the insurance policy, insurance application, surrender, and claim settlement to ensure that business operation risks are controllable and reduce losses arising from risk events. Meanwhile, it can render customers with a full range of services by selling related warehousing insurance products.

7. Pledge process management subsystem

It includes pledge review, inspection management, release management, and preservation management. It digitalizes and intelligentizes review management, control, release, and preservation of the goods in the pledge process, thus reducing the loss of goods and improving the pledge level. This ensure the stable and efficient development of the warehouse receipt pledge business.

8. Price monitoring and smart early warning subsystem

It includes price monitoring, price forecasting, and smart early warning. It monitors and issues smart early warnings of pledged goods' market situation and commercial prices in a real-time manner, improves business risk control capabilities, and reduces losses due to market fluctuations. Therefore, the commercial interests of lending institutions and loan applicants are safeguarded and there is a data basis for analysis.

9. Statistical analysis subsystem

It includes business analysis, market analysis, and report generation. By collecting and sorting financial logistics data, data is analyzed in combination with the development needs of companies. And the analysis results are fed back to users as reports and statements to help companies figure out operational levels, grasp market and customer needs, and improve service levels.

6.6 Security management and emergency support system

6.6.1 Construction requirements and significance

1. Construction requirements

With the wide application of the mobile Internet and big data, the risks and contradictions of nature and society are intertwined, and information security is gaining more weight. In the daily operation of transportation, warehousing, and distribution management, and other value-added services, security and emergency support are what companies must attach great importance to. A complete security management system is urgently required – if an emergency strikes, response can be quickly initiated, meaning a contingency plan is in place to minimize damage and ensure the security and controllability of the smart platform.

2. Significance

The security management and emergency support system can formulate safe and effective approaches for comprehensive transportation, warehousing supervision, and intensive distribution to smoothen the development of various businesses. It is an important system support and safeguard method; in the meantime, when an emergency strikes, the system will swiftly receive related information and notify relevant management personnel immediately, thus activiting the emergency treatment procedure. It tracks the handling process and displays it immediately, and uses the most effective means to quickly resolve emergencies and improve emergency response capabilities.

6.6.2 Business process analysis

The security management and emergency support system has collected basic information such as warehousing, transportation, and distribution. Input by business management personnel and related monitoring equipment forms a database after analysis, and enables a comprehensive inquiry function. The subsystems of security evaluation and security early warning constantly amend contingency plans to secure safety. The business process is shown in Figure 6-12.

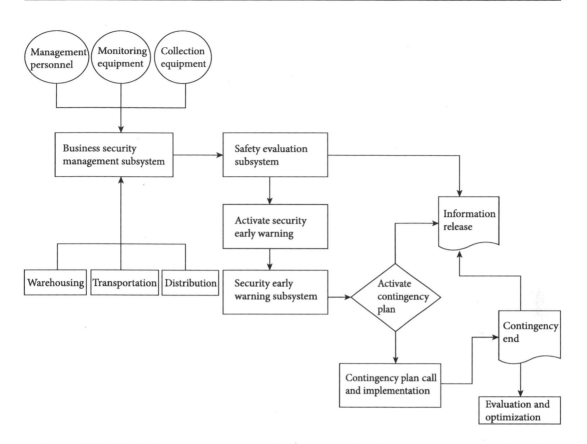

Figure 6-12   The business process of security management and emergency support system of a logistics company

6.6.3 Overall structure and function description

Targeted at the data and system security of the smart platform, a security management and emergency support system is set up to improve the business management system. Once an emergency occurs, the contingency plan will be immediately implemented at the fastest response. Specifically, it includes business security management, security evaluation, security early warning, and contingency plan subsystems. This is shown in Figure 6-13.

1. Business security management subsystem

It has three management modules: business infrastructure, business process security, and business personnel security. It mainly manages business processes and business personnel security access, thus meeting companies' security management requirements in transportation, distribution, and warehousing, ensuring their safe and reliable operation.

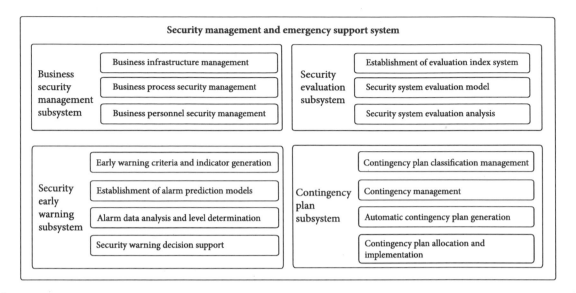

Figure 6-13   The main functional architecture of the security management and emergency support system

2. Security evaluation subsystem

It includes the establishment of the evaluation index system, security evaluation model, and security system evaluation analysis. It is used to evaluate security and statistical information, provide corresponding index data for security early warning, and guide companies to issue security early warning and activate contingency plans.

3. Security early warning subsystem

It generates early warning criteria and indicators, sets up alarm prediction models, analyzes alarm data, determines its level, and supports security early warning decision-making. Based on alarm source types and alarm analysis models, the security status of transportation, distribution, and warehousing of a company is analyzed to determine the type and level of alarms; alarms are then issued per the analysis results.

4. Contingency plan subsystem

It includes contingency plan classification, management, automatic generation, and allocation and implementation. The subsystem can determine whether to activate the contingency plan according to the alarm level, assist relevant personnel to implement contingencies through decision support and conduct an accident assessment, and input the successfully implemented countermeasures to improve the speed and quality of emergency response.

6.7 Big data application service system

6.7.1 Construction requirements and significance

1. Construction requirements

A large amount of data is generated in the operation of an company – especially in the business links of transportation, warehousing, loading and unloading, distribution, and logistics finance. This requires support from advanced technologies to summarize, classify, and analyze the data generated so that potential laws hidden behind the data are explored to benefit enterprise analysis, prediction, and decision-making.

2. Significance

The big data application service system can allow many systems to interact, coordinate a large amount of information, mine the information, and provide enterprise with in-depth business distribution and operation level analysis. This helps companies understand customers' market strategies, supply chain operations, and sales strategies, thus they can design targeted and personalized services; Simultaneously, by collecting, analyzing and processing business operation data, companies can understand their own business operations and development trends, profit levels of various businesses, growth rates, market demand, and new business demand direction. It can help those in charge to adjust development strategies and decision-making in a timely manner for diversified development goals of low cost, high efficiency, quality service, and environmental protection.

6.7.2 Business process analysis

The big data application service system mainly categorizes, extracts, and converts related business data according to a company's individual business development, and displays the results in visual diagrams for business reporting. This is shown in Figure 6-14.

6.7.3 Overall structure and function description

According to actual business needs, a big data application service system is designed with the purpose of classifying, extracting, collecting, integrating, and sharing data from various business systems through cloud computing, data warehousing, data mining and GIS. It mainly analyzes and displays business in a digital and graphical fashion, provides management personnel with display reports, business evaluation, and auxiliary decision-making services, and upstream and downstream companies with big data analysis service. This system involves data classification and summary, statistical and predictive analysis, operational analysis, and business intelligence. The main functional architecture is shown in Figure 6-15.

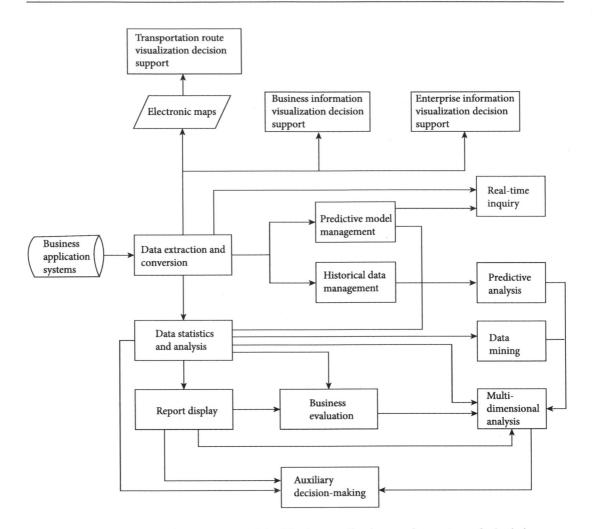

Figure 6-14   The business process of the big data application service system of a logistics company

1. Data classification and summary subsystem

It includes four data analysis modules: logistics transportation, logistics warehousing, logistics distribution, and other related data analysis. It is often used for classification and preliminary processing of data generated. A comprehensive analysis makes complex data orderly and lays a solid foundation for subsequent statistical analysis.

2. Statistical analysis subsystem

It includes three modules: data calculation, display reports, and business evaluation. The data source is the classification and summary subsystem. After collection, filtering, and sorting, business is digitally and graphically analyzed through the three modules, and a data base is layed for other related businesses.

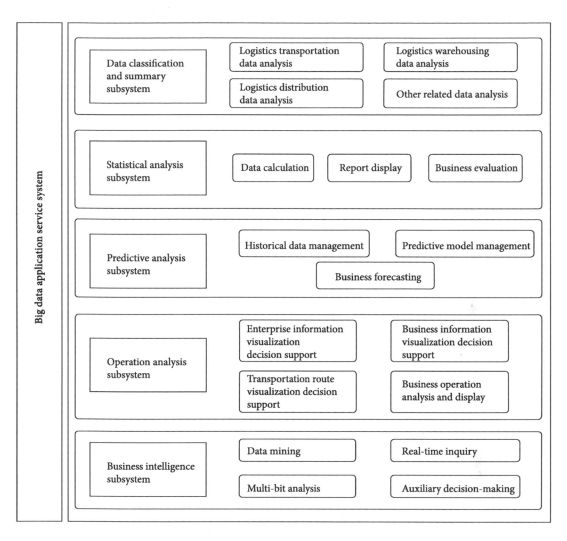

Figure 6-15 The main functional architecture of the big data application service system

3. Predictive analysis subsystem

It includes three modules: historical data management, predictive model management, and business forecasting. It is normally used for historical data management, predictive model management, and predictive analysis. That is, corresponding predictive analysis models are designed and built according to a user's business development and operating data. It is able to predict the development trend of core, auxiliary, and value-added business.

4. Operation analysis subsystem

It includes enterprise decision support for business information visualization, transportation route visualization, and business operation analysis and display. It combines GIS and electronic maps to visualize users' enterprise information, and present the analysis results through diagrams.

5. Business intelligence subsystem

It includes four modules: data mining, real-time inquiry, multi-bit analysis, and auxiliary decision-making. That is, via these technologies and methods, business data is transformed into information with commercial value to improve the intelligence of core, auxiliary, and value-added business analysis.

CHAPTER 7

BULK COMMODITY SUPPLY CHAIN INFORMATION SERVICE PLATFORM

7.1 Overview

Bulk commodities refer to the basic raw materials that participate in commodity circulation and are used for industrial and agricultural production and consumption. Compared with other commodities, they have huge transaction volumes, large price fluctuations, and are easily classified and standardized. The supply chain is an organization form that, guided by customer needs, aims to improve quality and efficiency by integrating resources so that product design, procurement, production, sales, and service processes are efficient and collaborative. The bulk commodity supply chain revolves around core enterprise and transports bulk commodities to downstream businesses by controlling business flow, information flow, logistics, and fund flow from the procurement of raw materials to the making of intermediate and final products. Suppliers, manufacturers, distributors, retailers and end users are therefore connected in a complete functional network chain structure.

The informatization of bulk commodity supply chain services is inevitable in its development. In response to the present contradiction between the development status and demand in the Chinese market, it is urgent to adopt modern management techniques and concepts to reengineer business processes, intensify control, and establish a scientific and reasonable bulk commodity supply chain service platform.

As per the Guiding Opinions on Actively Promoting Supply Chain Innovation and Application issued by the State Council, it is necessary to build a comprehensive, informatized, and coordinated bulk commodity supply chain information service platform. This can coordinate, servitize, and smarten the supply chain, improve circulation modernization, and actively develop supply chain finance, thus, unified and scientific decision-making is achieved among enterprise at all nodes of

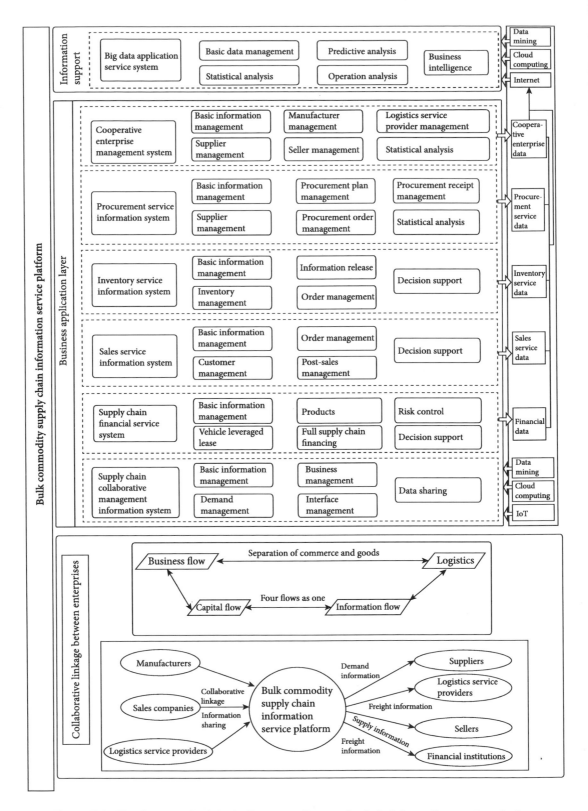

Figure 7-1 The framework of the bulk commodity supply chain information service platform

the supply chain, operation and management is enhanced, and the overall development level of the Chinese supply chain is improved. This framework is shown in Figure 7-1.

This platform is composed of three modules: information support, business application layer, and inter-enterprise collaboration and interaction. It solves untimely, incomplete, and inaccurate information flow problems between nodes and effectively ensures balance and coordination between them. This promotes data openness and transparency, which contributes to planning logistics tasks as a whole, ensures high quality services, and coordinates the relationship between service integrators and providers, which greatly propels propels logistics development.

7.2 Cooperative enterprise management system

7.2.1 Demand analysis

1. Background

Based on the needs of customer management, a cooperative enterprise management system is established for node enterprises involved. This achieves sort management of the node enterprises and improves customer management efficiency. Each logistics company is, in effect, the system manager while the node enterprises in the bulk commodity supply chain are the users. For example, a coal production company supplies coal power and chemical companies, but also produces raw coal diggings.

2. Significance

The construction of the cooperative enterprise management system can effectively integrate various cooperative enterprises in the supply chain, perform sort management, and provide strong support for the comprehensive management and control of the industrial chain by government departments. In addition, it can also greately promote connection and cooperation between logistics companies and the various node enterprises, enhance information sharing, ensure cooperation and trust, and render full supply chain services.

The cooperative enterprise management system is the foundation of the bulk platform. It both promotes the informatization and intelligent management of various cooperative enterprises, bulk commodity logistics service providers, and downstream market customers, and supports the subsequent systems with basic information.

3. Functional requirements

The cooperative enterprise management system is required to perform sort management on the service objects in the bulk system, including basic information management of suppliers, manufacturers, sellers, and logistics service providers. Simultaneously, it should evaluate customer value at all stages, so that different management approaches are applied to customers of different values.

7.2.2 Business process and data flow analysis

1. Business process

The cooperative enterprise management system has to collect information from node enterprises, bulk commodity logistics service providers, downstream market customers and their related businesses, apply customer feedback, and perform statistical analysis of various basic information. This process is shown in Figure 7-2.

Cooperative enterprise in the bulk commodity supply chain are suppliers, manufacturers, sellers, and logistics service providers, among whom there are business relationships. Sellers makes purchase requests to manufacturers, who then makes a purchase from the supplier; the logistics business then delivers raw materials to the manufacturer, who then ships the product to the seller; the shippment is taken care of by the logistics service provider.

2. Data flow

The cooperative enterprise management system has found a system database to collect and input supplier information, manufacturers, sellers and logistics service providers, and generate final data via processing. This is shown in Figure 7-3.

It takes the cooperative enterprise management database as the core and extracts basic data such as customer, product, and credit evaluation information from the four management subsystems. Based on the extracted data, a statistical analysis about the cooperative enterprise is run.

7.2.3 Overall structure

Based on the bulk supply chain business needs of logistics companies, a cooperative enterprise management system is founded. It is combined with the service requirements of the four types of cooperative enterprise: suppliers, manufacturers, sellers, and logistics service providers. It improves customer management efficiency and has six subsystems: basic information management, supplier management, manufacturer management, seller management, logistics service provider management, and statistical analysis. Its overall structure is shown in Figure 7-4.

The cooperative enterprise management system is designed in line with the use and management needs of system and related supply chain users. It collects and manages their information and provides basic support for normal system operation; the supplier, manufacturer, seller, and logistics provider management subsystems enable customer information enquiries, and design corresponding functional modules based on the needs of different users; statistical analysis subsystems conducts multi-angle analysis and generates corresponding reports for managers to make a decision.

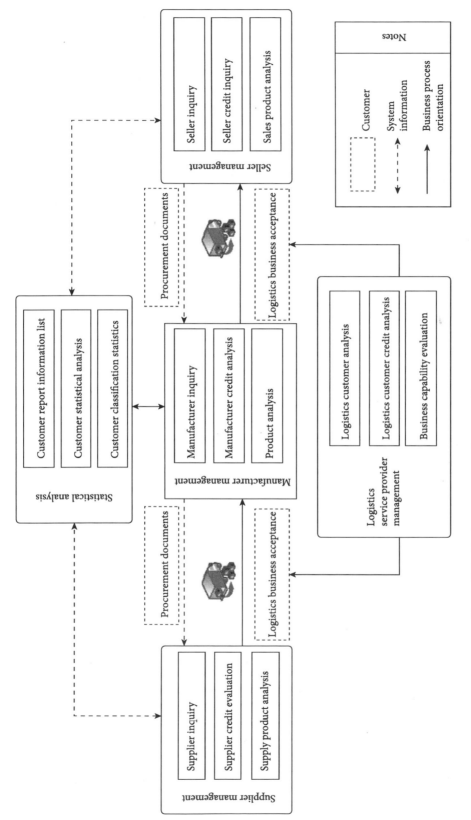

Figure 7-2　Business process of cooperative enterprise management system

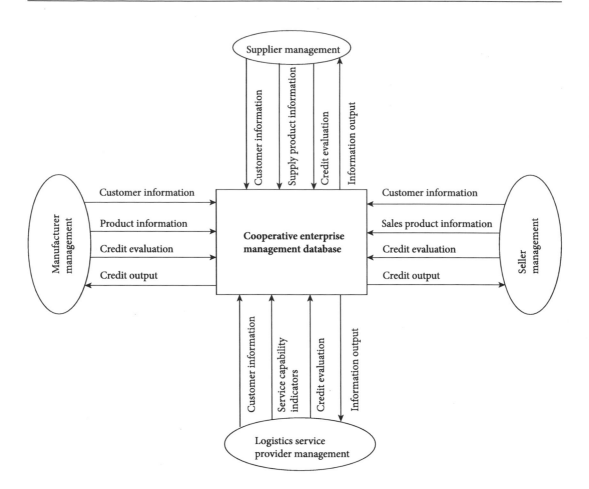

Figure 7-3   Data flow of the cooperative enterprise management system

7.3 Procurement service information system

7.3.1 Demand analysis

1. Background

The procurement department is a crucial entry point for materials and in close contact with other departments. It keeps business contact with the production, finance, and inventory departments to draw procurement plans based on production and material requirement forecasts, and to complete a series of purchasing operations.

Based on the demand for procurement management in the construction of the bulk service platform, the procurement service information system is established for users. Traditional and new information technogloies are comprehensively applied to perform two-way control and tracking of the entire process and improve the management of material supply information.

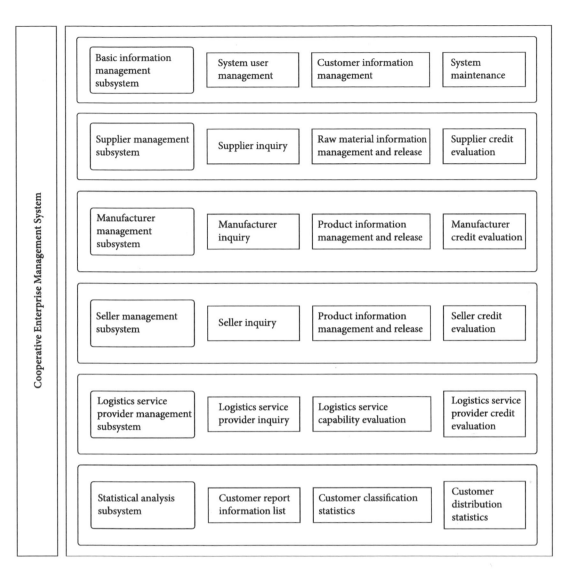

Figure 7-4 Overall structure of the cooperative enterprise management system

2. Significance

Procurement management is the center of bulk commodity supply chain management and an important part of the entire management system. It is the first step of bulk commodity production and operation activities. When it is compared with traditional management approaches, the establishment of a procurement service information system can place and execute purchase orders faster and more accurately, summarize the reception, inspection, and warehousing of goods, optimize the procurement process by rigorously monitoring all links, and improve efficiency.

The founding of the bulk procurement service information system can understand supplier and customer information punctually, avoid cost losses arising from information asymmetry, ensure real-time and accurate information in both supply and demand, and offer strong support for procurement efficiency and sales cost reduction. Meanwhile, with supply chain coordination, it can make reasonable use of upstream material resources, liquidity, and production and operation facilities and venues. This solves problems such as tight funds and insufficient facilities; such procurement plays a pivotal role in solving all sorts of problems and improving economic interests.

3. Functional requirements

As the initial step of production and operation activities, procurement has a great impact on supply, production, and sales. Logistics procurement and organization of supply are the basis of industrial and commercial business activities. They play a decisive role in ensuring their continuity. The procurement service information system ought to cooperate with inventory and sales services to effectively manage and control procurement processes, such as entering contracts, bill management, the inspection and warehousing of purchased goods, and payments – creating a complete system.

7.3.2 Business process and data flow analysis

1. Business process

The complete procurement process consists of procurement demand planning, procurement application, selection of suppliers, determination of suitable prices, order contracts, order tracking, replenishment control, inspection of goods and warehousing, appropriation of funds, return processing, closing and file maintenance, etc. It involves production, inventory, procurement, auditing, and finance departments. This process is shown in Figure 7-5.

1) Procurement demand planning and procurement application

The procurement department draws, based on the inventory and outsourcing lists provided, a procurement plan which should be submitted to the review department. Once approved, the department can submit a procurement application. When that is passed, it is moved on to the next step: selection of suppliers; if that fails, it is returned to the buyer for adjustment.

2) Selection of suppliers

There are three factors to consider when selecting and confirming suppliers: price, quality and delivery time. As suppliers are of vital importance to companies, establishing and developing a relationship with them is a critical step of business strategy, especially in JIT (just-in-time) production modes, where higher requirements are found for the stability and reliability of the partnership.

3) Determination of a suitable price

Once potential suppliers are sifted out, price negotiations take place. Generally, the price that suppliers offer is relatively stable in such environments.

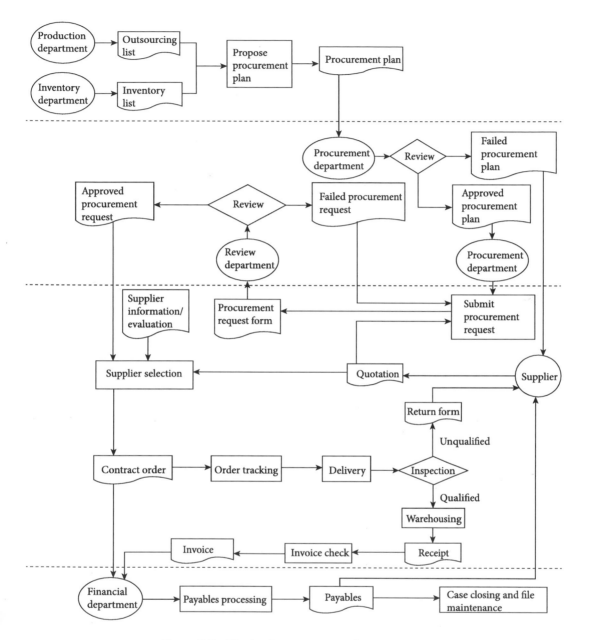

Figure 7-5 The business process of procurement

4) Order contract

This is the procedure of placing the order and entering the contract. The order or contract is legally binding in which the requirements, rights, and obligations of the buyer and the seller must be specified.

5) Order tracking, replenishment control

The procurement department is supposed to urge suppliers to deliver on time. Buyers not only have to supervise this but also promptly detect problems in the process and take measures. It is also responsible for addressing any change in delivery requirements and negotiations with the supplier.

6) Inspection of goods and warehousing

Once the goods arrive, the buyer shall supervise relevant personnel to check them to ensure their quality and quantity is consistent with the order requirements. If necessary, they shall identify damage in goods, and finally notify the finance department to settle the payment.

7) Appropriation of funds

The finance department checks the procurement order, receipt and invoice, and makes the goods payment. When the supplier has passed the inspection and acceptance, invoices will be issued; when the payment is required to be cleared, the content of the invoices should be checked by the procurement department first before the finance department initiates the payment procedures.

2. Data flow

Based on analysis of the business process and relevant procurement content of each node, data processing of the procurement business is investigated from the perspective of data flow. The data flow between the procurement service subsystems is combined to design a data flow of the procurement service, as shown in Figure 7-6.

(1) A procurement plan is drawn based on the inventory and outsourcing lists provided by the production department. It is then submitted for review; once approved, it is recorded in the procurement plan document; if denied, it is returned.

(2) According to the materials required in the procurement plan, suppliers are selected and the purchase order is generated. First, a quotation form is prepared as per the quotations of each supplier; then the best supplier is selected with reference to their evaluation records; next, contact is confirmed according to basic supplier information; lastly, a purchase order is generated and sent for review; once passed, a contract is generated and the purchase order document is archived; the contract is mailed to the supplier.

(3) The supplier ships the goods; the consignee checks and modifies purchase order documents and the quantity, and issues an inspection form to notify relevant departments. Subsequently, the results are input into the supplier evaluation record (unqualified product record). If there is a unqualified product that needs to be returned, the order document, purchase order document and the quantity of goods are modified.

(4) The finance department clears payment for the supplier according to the qualified order while checking invoices issued by the supplier.

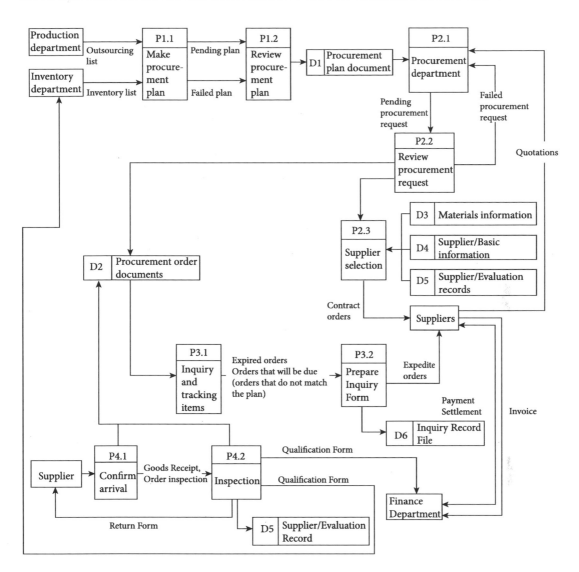

Figure 7-6 The data flow of procurement service

Note: P represents related procurement data processing departments while D is procurement data information.

7.3.3 Overall structure

According to the management business needs of the bulk commodity supply chain, a procurement service information system is established which combines the service requirements of three types of customers: suppliers, manufacturers, and logistics service providers. Demand and supply information is therefore shared and the related business of the procurement process managed. It embraces six subsystems: basic information management, supplier management, procurement plan management, procurement order management, procurement acceptance management, and statistics analysis. It is shown in Figure 7-7.

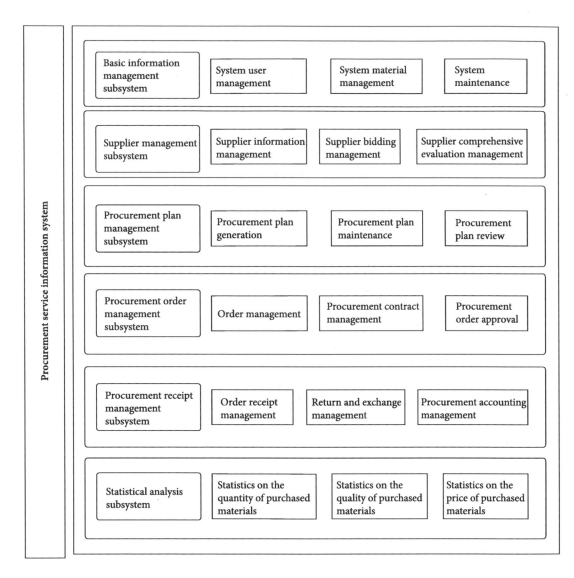

Figure 7-7   Overall structure of the procurement service information system

This system is designed in line with the use and management needs of system users and related suppliers. The basic information management subsystem handles user information and purchased materials while supporting normal system operation. The supplier management subsystem performs corresponding management of core and bidding suppliers. The plan management, order management subsystem, and acceptance subsystem manages specific procurement steps; their design corresponds to functional modules based on different user needs. The statistical analysis subsystem conducts multi-angle analysis and generates corresponding reports and statistical charts for decision-makers.

7.4 Inventory service information system

7.4.1 Demand analysis

1. Background

To meet changes in customer demand, enhance the flexibility of production plans, and overcome fluctuations in raw material delivery teams, upstream and downstream node enterprises in the supply chain must stock a certain amount of raw materials and finished products. However, in the industrial supply chain, the inventory management of some enterprise is simple and extensive without the support of informatization and data technology – it has not helped with decision-making and operation as it should. Therefore, the inventory service information system is added to the bulk commodity platform. This rationalizes inventories in terms of cost and service response speed.

2. Significance

The inventory service information system can integrate all company inventories into the supply chain. It establishes a unified platform to reduce the overall stock level in the, to optimize inventory management, and effectively control all supply chain links. This ensures the best flow process from raw materials to products, and provides an informationized and efficient management platform. All node enterprises in the supply chain can depend it on it to manage inventory business. Through data mining, information interconnection, and IoT, it collects relevant information and releases relevant decision-making information too.

3. Functional requirements

This system adopts modern information and computer technology to integrate the actual needs of the supply chain inventory management, systematically collect, process, transmit, store and exchange inventory information, and provide corresponding information as users require. As a result, it visualizes and informatizes supply chain inventory management, provides macro decision support, and effectively manages inventory through collaboration.

7.4.2 Business process and data flow design

1. Business process

Inventory services include inventory management, warehouse management, and order management. It involves the production, procurement, supply, and finance departments. This is shown in Figure 7-8.

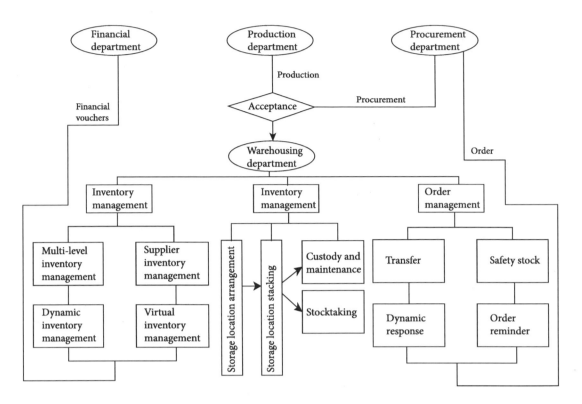

Figure 7-8   The business process of inventory services

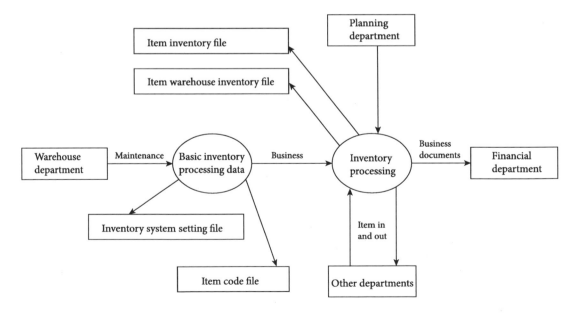

Figure 7-9   The data flow of inventory services

Multi-level inventory management is the center of inventory management: the multi-level inventory in the supply chain refers to the existing inventory of a certain node enterprise. Multi-level inventory management, via information sharing, makes inventory control strategies to optimize costs throughout the supply chain. Stock management is about goods allocation arrangements, goods allocation stacking, custody and maintenance, and stocktaking. Order management includes allocating and warehouse shifting, safety stock, dynamic response, and order reminders. Among these, allocation and warehouse shifting allocate goods between different warehouses to meet market demand. When the goods reach the safety stock, the inventory department makes replenishment requests.

2. Data flow

According to the relevant inventory business content, the data processing mode of the actual inventory business is investigated from the perspective of data flow. And the flow of data information between inventory service subsystems is combined to design the data flow of inventory services, as shown in Figure 7-9.

Data flows from each planning department to the warehouse department, which properly arranges the inventory materials according to the data transmitted. At the same time, it should process the inventory as required by the business department and update the information of stored items. In addition, each node enterprise of the supply chain inquires about inventory information through authority. When the order point is reached, the warehouse department issues a notice on the platform to replenish the goods.

7.4.3 Overall structure

The subsystems of the inventory service information system consist of basic information, inventory, order, information release, and decision support. The structure is shown in Figure 7-10.

The inventory service information system adopts modern information technology, IoT and big data mining to integrate warehousing and inventory management requirements of all supply chain node enterprises to efficiently transmit and share inventory information throughout the supply chain. In the meantime, to integrate this and enable inventory information to circulate freely between all supply chain users, it integrates inventory management of all supply chain node enterprises. Finally, it contributes to reducing the number of production preparations, overcoming fluctuations in raw material delivery times, and enhances production planning flexibility.

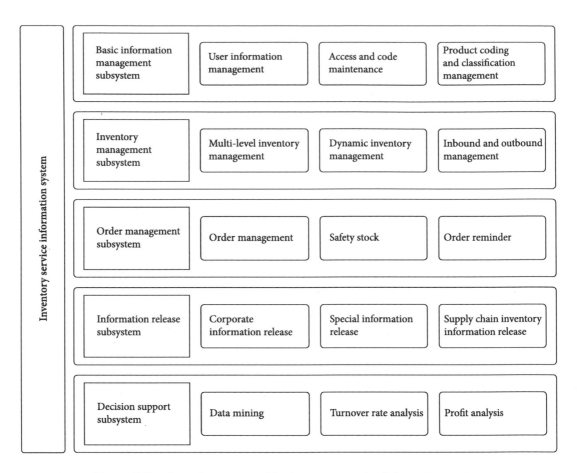

Figure 7-10   Overall structure of the inventory service information system

7.5 Sales service information system

7.5.1 Demand analysis

1. Background

The central business activity of a company is product sales. The ultimate production and operation objective is to profit through sales. Therefore, sales services are the decisive factor for companies to maximize business interests. However, in the supply chain that logistics companies serve, some node enterprises lack the ability to analyze systematically nor the necessary basic data support, which have led to deficiencies in sales information management. Based on the demand for sales service information in the supply chain, modern methods like data mining are employed to build a sales service information system that enables logistics companies to analyze and process various sales information. This utilizes a more scientific and comprehensive basis for decision-making. Information is also integrated and data shared throughout business operations to innovate and optimize the enterprise value chain.

2. Significance

The sales service information system covers multi-party information of suppliers, distributors and other business partners in the supply chain, integrates information and fund flows, and optimizes logistics across multiple departments. It improves the response speed of the sales service system and customer satisfaction. Simultaneously, it shares sales information among supply chain users, and adopts data mining technology to enable platform users to grasp and track real time dynamic information of their products, thus forming a rapid response mechanism and ultimately improving sales management.

3. Functional requirements

In order to enhance the sales level of the supply chain, big data technology, information transmission and processing technology are used to enable their sales departments to grasp and track real-time supply chain product demand and transaction information. Statistics and sales data analysis provide reliable information support via collaboration between subsystems.

7.5.2 Business process and data process design

1. Business process

The principal business of the supply chain that logistics serves comprises of customer, order, and after-sales management. It covers suppliers, distributors, logistics providers and other business partners. This is shown in Figure 7-11.

Statistical analysts employ a variety of data approach methods to analyze the needs of potential customers and prepare reports. Sales personnel of each supply chain node can inquire about the analysis results. After customers inquire or actively quote, a sales plan and market supply and demand are combined to determine a reasonable price and a business negotiation takes place. Once an agreement is reached, the contract is entered and approved, and the system automatically records the contract ledger. Meanwhile, a shipping plan, as stipulated by the shipping terms in the contract, is formulated with the winning logistics provider to ensure punctual shipment. After confirming the acceptance of goods, after-sales tracking takes place. Any return or exchange of goods will be processed in a timely manner. If the product passes inspection and acceptance, receipts and payments will be made according to the financial accounting period.

2. Data flow

According to the sales business content of supply chain users, the actual sales business data processing model is investigated from a data flow perspective. The flow of data information between the sales service subsystems is combined to design this system, as shown in Figure 7-12.

This revolves around user, order, commodity, and after-sales management. The sales service information system managers divide responsibilities of the sales department of supply chain node users, code the products for sale, and update information. With this system, users can inquire about supply and demand, and receive data support for sales decisions through analysis reports available.

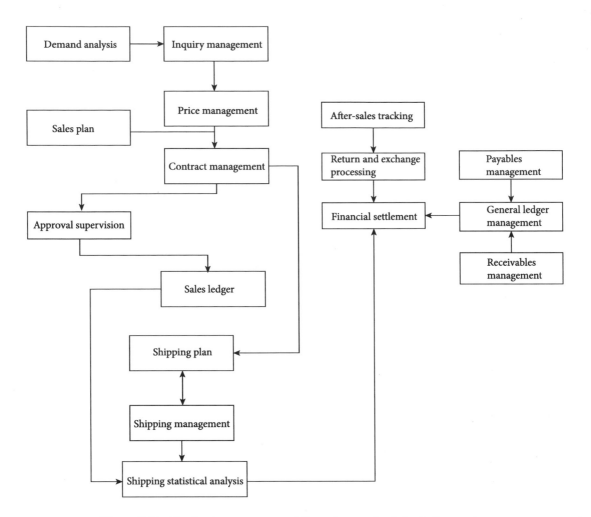

Figure 7-11 The business process of the sales service information system

Regarding after-sales management data processes, the system feeds back the offline return and exchange voucher information to the sales service system to generate online data.

7.5.3 Overall structure

This has five subsystems: basic information management, customer management, order management, after-sales management, and decision support, as shown in Figure 7-13.

It applies integration technology to coordinate the operation of various subsystems. The basic information subsystem regulates basic transaction information through user information management, access and password settings, and product coding management. The customer management subsystem analyzes customers to capture their prior needs and improv economic interests of enterprise. The order management subsystem associates the sales plan with the sales contract, guides the implementation of the latter through the former, and regulates supervision of

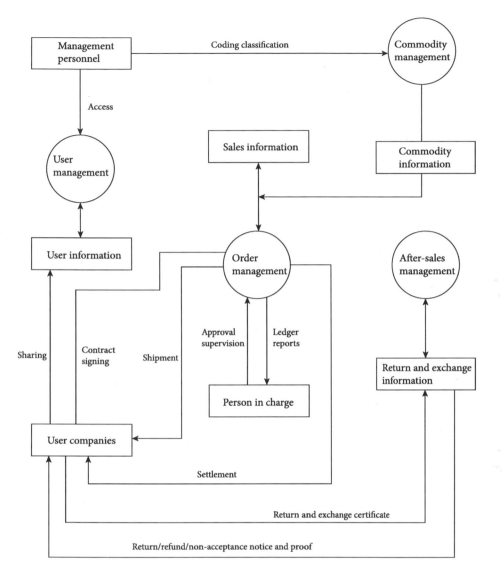

Figure 7-12   The data flow of sales service information system

contract signing and execution. The after-sales subsystem improves user satisfaction via after-sales tracking, return and exchange. The decision support subsystem employs data mining and cloud computing to analyze sales data for decision-making.

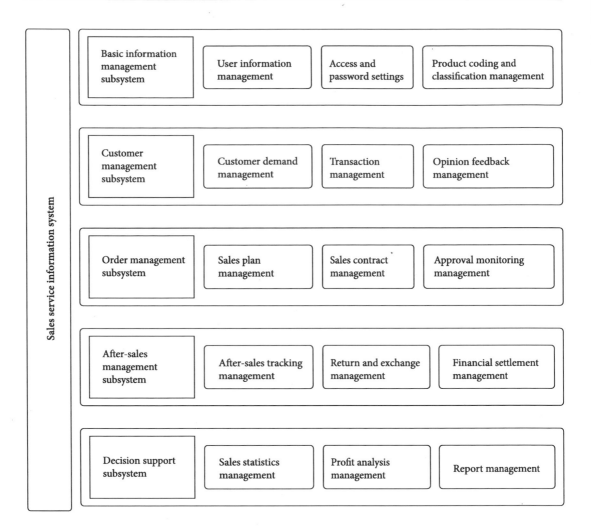

Figure 7-13   The overall structure of the sales service information system

7.6 Supply chain financial service system

7.6.1 Demand analysis

1. Background

Owing to the high financing costs and difficulties for supply chain users in logistics, a financial service system is established to provide strong support for solving financial problems.

It extracts information from different subsystems according to established rules, and mines, processes, and integrates shared supply chain financial data within the platform. For higher operational efficiency, financing management, financial pledge supervision, and risk management are combined. Advanced technologies are also applied to production and sales companies that are served by the supply chain logistics, and financial institutions.

2. Significance

The significance of this system lies in the enhancement of core competitiveness of the supply chain, the promotion of effective interaction between finance and the industrial economy, the combination of the financing needs of supply, production, and sales, and the connection between core users and upstream and downstream users.

The supply chain financial service system can provide different levels of information and support decision-making for different customers, meet the needs of different platform users, and collect, process, organize, store, release, and share supply chain financial information. Thus, costs are reduced and efficiency is improved.

3. Functional requirements

Functionally, the system is able to help users accomplish financing and credit, guide standardized business processes, strengthen the collaborative relationship with financial institutions, deepen collaborations, improve risk management capabilities and service levels, increase customer satisfaction, and form competitive advantage. Meanwhile, it is necessary to ensure that financing risks are controllable, that financial services co-develop, that financing difficulties are eliminated, and that profitability is enhanced.

7.6.2 Business process and data process design

1. Business process

To improve operational efficiency and asset utilization of the supply chain, the characteristics of the operation and management cycle are combined and targeted at the characteristics of the capital gap in operation. The differences in risk points of the borrower's financing demand are divided into

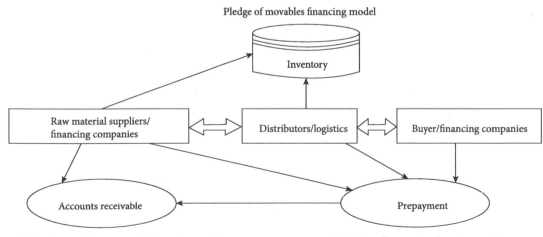

Figure 7-14 The business model of the supply chain financial service system

three categories: accounts receivable financing, movable property pledge financing, and accounts payable financing. This is shown in Figure 7-14.

1) Supply chain finance in the procurement stage – collection on delivery

The financing mode of collection on delivery refers to the service model where logistics companies make a certain percentage of the payment to the supplier (manufacturer) in advance. They obtain the transportation agency right of the goods as agreed, while the agent supplier collects payment and the buyer pays the goods to the logistics companies in one go when picking up the goods, as shown in Figure 7-15.

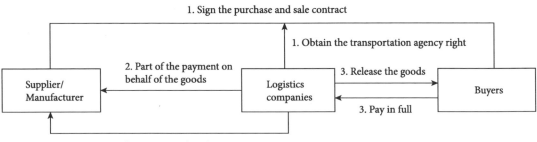

Figure 7-15   The business process of collection on delivery

As per the purchase and sale contract, logistics companies obtains the right to carry the goods. They make a certain percentage of prepayment for the buyer to obtain the ownership of the pledged goods. The buyer makes the full payment to obtain the goods and then logistics companies make the remaining payment, with service charge deducted to the supplier. Under this financing mode, in addition to interests from traditional logistics links such as cargo transportation, companies also gain interest-free capital due to delayed payment – this can be used for additional profits.

2) Supply chain finance in the inventory stage – movables pledge

The financing mode of movables pledge refers to the financing business in which companies pledge their inventory and use the income generated as the first source of repayment. In this mode, financial institutions will enter a guarantee contract with financing companies or pledged property repurchase agreements, as shown in Figure 7-16.

When a financing company applies for movables pledge financing, it has to hand over legally owned goods to the storage supervisory party the bank approves, during which only the property in goods is transferred, not the ownership. Once the goods are shipped, the bank will finance them in a certain proportion according to the condition of the goods. When the consignee pays the bank, it gives delivery instructions and hands over the property. If the consignee fails to repay the bank within a specified period, it can auction the goods in both domestic and international markets.

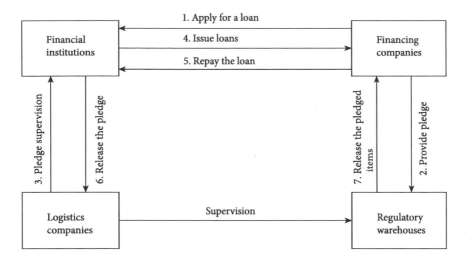

Figure 7-16 The business process of the movables pledge

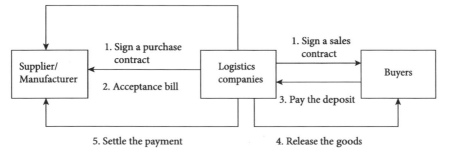

Figure 7-17 The business process of substitute selling service

3) Supply chain finance at the sales stage – substitute selling service

Logistics companies replace the supplier (manufacturer) to sell to the buyer and obtain ownership of the goods; they transport the goods to the designated warehouses and the buyer pays a deposit to acquire the corresponding quantity of goods; when all the goods are released, payment is cleared. This is shown in Figure 7-17.

As agreed, logistics companies will replace suppliers to sell goods to the buyer, enter the purchase and sales contracts with the buyer, and obtain goods ownership. They pay the supplier for goods generally though commercial acceptance bills – the buyer pays a certain amount as a deposit and they release the corresponding goods and clear the payment with the buyer. The substitute selling service allows logistics companies to gain business income from cargo transportation, warehousing, and distribution processing and earn capital gains through the price spread of goods. It also strengthens cooperation between the supply chain.

2. Data flow

According to the financial business content of supply chain node users, the data processing mode of the actual supply chain is investigated from the perspective of data flow. This data flow between service subsystems is combined to design the financial service system, as shown in Figure 7-18.

This flows among manufacturers (suppliers), logistics providers, distributors, end users, and banks. It contributes to bill, post-loan, loan, and margin management. Lastly, all data is aggregated to the processing center by the information platform for supply chain node users.

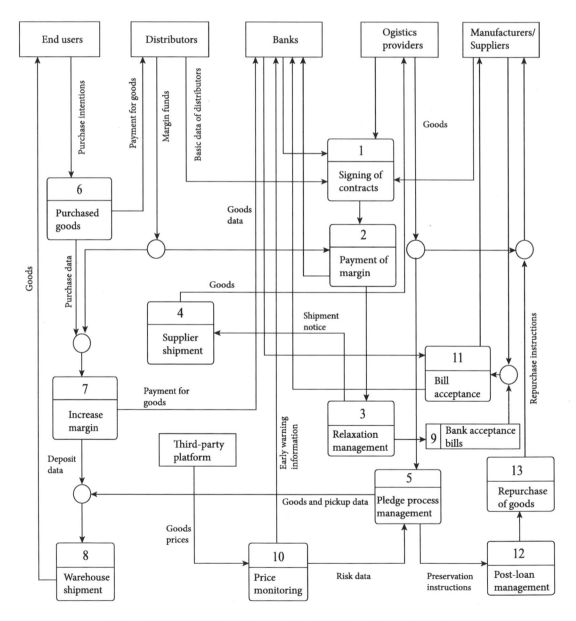

Figure 7-18   Data flow of the supply chain financial service system

7.6.3 Overall structure

Based on analysis of the supply chain financial service business and the current financial status of the supply chain users, the system is designed. It includes basic information, vehicle leveraged lease, product, supply chain financing, risk control, and decision support subsystems, as shown in Figure 7-19.

This system concentrates on logistics and relies on the abundant guarantee of supply chain resources; it solves financing difficulties of node users and has much development potential. As a cross-industry business information platform, it covers business marketing, full logistics monitoring, scheduling and optimization, data collection and processing, risk identification,

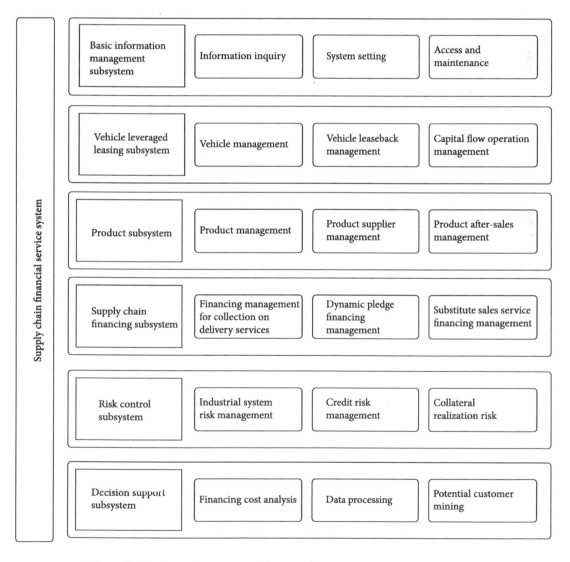

Figure 7-19 Overall structure of the supply chain financial service system

evaluation and control, and clearance support. These functions and supporting operation mechanisms eventually evolve a management information platform based on the tripartite of banks, customers, and logistics companies. They support the strategic development of supply chain finance, standardize and optimize operation, simplify financial clearance, reduces financing costs, and improves core competitiveness.

7.7 Supply chain collaborative management information system

7.7.1 Demand analysis

1. Background

The higher the degree of business synergy in the supply chain system, the greater the functions and effects are. In a general supply chain system, core enterprises are scattered in different places and links, and their individual advantages, value-added strength and development goals are difficult to coordinate and unify. The establishment of this system can manage business collaboration between users at each node by clarifying the identity of each role, setting up common standards and strategies, and determining the collaborative management mode and specific work.

Logistics companies are responsible for the management of business collaborations on multiple supply chains. This system can unite independent enterprises that are scattered in different regions and value-added links are provided. This can maximize the strength of node enterprises in the bulk cargo supply chain system.

2. Significance

The construction of the supply chain collaborative management information system improves the degree of synergy of the supply chain. Each core user integrates resources through advanced information technology and management methods for coordination and synchronous operation of all nodes. Thus, scattered independent enterprise in the supply chain is united, contributing to optimal performance.

All logistics companies should take collaborative mechanisms as the premise, collaborative technologies as the support, and information sharing as the basis. From the systematic big-picture outlook, it promotes the collaborative development of the supply chain – both internal and external – and maximizes interests while improving competitiveness.

3. Functions

The functional goal of the system is to allay conflicts and internal friction, better divide labor, and cooperate through collaborative management. Supply chain node users must keep in mind mutual objectives and strive for common goals; a fair profit-sharing and risk-sharing mechanism must be established; extensive and deep cooperation must be conducted on the basis of trust, commitment, and flexible agreement; an IT-based information and knowledge sharing platform must be built for timely mutual communication; and business processes oriented at customers and collaborative operations must be reengineered. This system improves the competitiveness of the supply chain by enabling rapid response.

7.7.2 Business process and data flow design

1. Business process

By nature, this is a consulting and decision-making organization that provides the best supply chain solutions for all participants in multiple industrial chains and proposes common strategies. It mainly engages in information integration and planning to enable node users to share information and collaborate on mutiple aspects, including user collaboration, demand management, business management, interface management, and data sharing. This is shown in Figure 7-20.

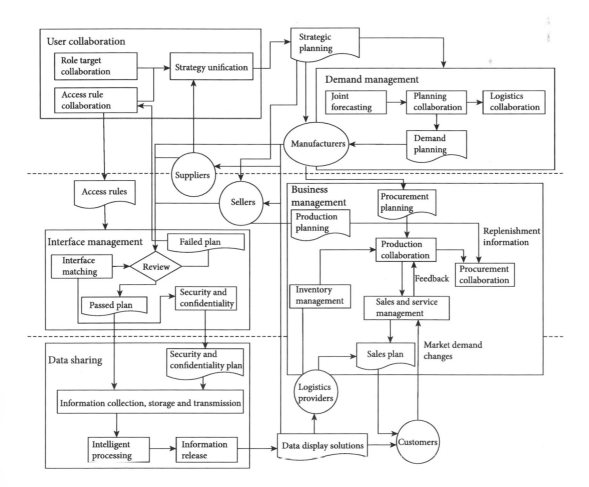

Figure 7-20   The business process of supply chain collaborative management information system

Through the collaborative management of all users, bulk commodity supply chain users determine and unify value goals and formulate a common alliance strategy. Suppliers, manufacturers, sellers, and logistics companies jointly make demand forecasts and determine collaborative plans, production and purchase, inventory, sales, and logistics and transportation. On top of that, through interface matching and security management, data is obtained and intelligently processed so that supply chain users can share inventory information and feedback from the external market. Internal and external users adopt the platform's communication and application function modules to participate in its business process, thus greatly simplifying integration and information network processes.

2. Data flow

By analyzing and sorting the collaborative management business of the supply chain and combining the data flow between subsystems, it will analyse collaborative work, targeted at suppliers, manufacturers, sellers, logistics providers and customers. This is shown in Figure 7-21.

In the bulk commodity supply chain, users are usually engaged in collaborative management of business activities. This includes strategic alliances, demand analysis, production plans, warehouse delivery, transportation plans, and purchases of goods.

7.7.3 Overall structure

As required by the supply chain collaborative management business, big data, IoT, and cloud computing are adopted to perform data analysis and establishe a supply chain collaborative management information system. This consists of five subsystem parts: basic information management, demand management, business management, interface management, and data sharing. This is shown in Figure 7-22.

The information system is designed by integrating the data requirements and usage of the supply chain system; basic information sets basic data and coordinates necessary alliance strategies; demand management informatizes, standardizes, and integrates demand management; business management coordinates operation product manufacturing, procurement, storage, sales, and service; interface management ensures the security and confidentiality of information sharing and data transmission via interconnection of interfaces within the system; the data sharing subsystem strengthens collaborative management and data information sharing as well as improving supply chain flexibility.

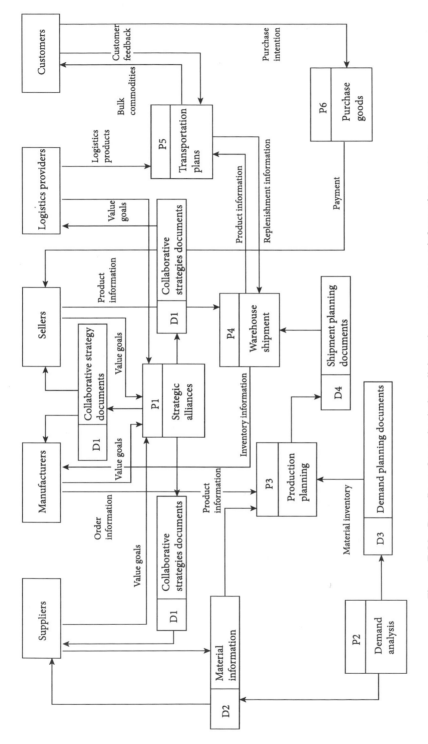

Figure 7-21 Data flow of supply chain collaborative management information system
Note: P stands for related data processing departments, D for data information.

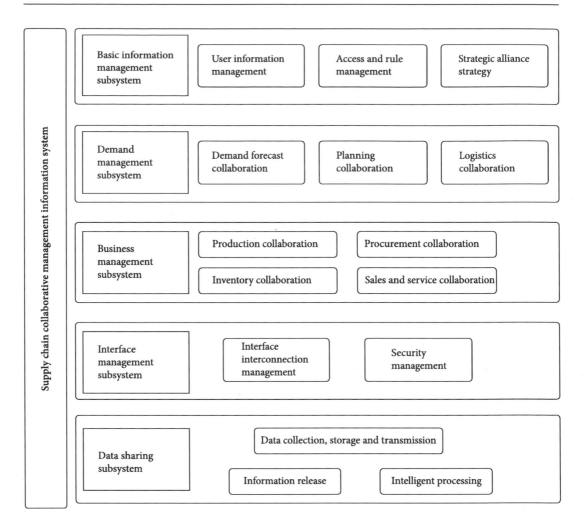

Figure 7-22   Overall structure of supply chain collaborative management information system

7.8 Big data application service system

7.8.1 Demand analysis

1. Background

As the core network chain of the logistics industry, the greater complexity of the supply chain will thoroughly alter market boundaries of enterprise, business portfolios, and business and operation models. Companies that manage the supply chain system and supply chain users must find better tools to quickly and efficiently maximize data value.

According to the business operation process of each user, a large amount of data information requires summary, analysis, and processing in bulk commodity supply chain collaborative management. Therefore, exploring potential laws hidden behind the data is of great necessity; it will play a critical role in the improvement of a company's decision-making and operational efficiency.

2. Significance

Establishing a big data application service system can effectively enhance the infrastructure, business operation, and management decision-making of user enterprises as well as the construction and application of supporting resources. It can also increase resource utilization, management level, and operational efficiency.

It combines the current business situation of logistics, targets the needs of platform applications, adopts advanced technologies, and collects, stores, and analyzes massive amounts of data. It also conducts data processing, data analysis and mining, and eventually obtains useful information to offer smart decision-making support for the operation and management of bulk commodity supply chain users.

3. Functions

The big data application service system, with support from advanced technologies, enables automatic sensing and identification, roots tracing, real-time response, intelligent decision-making, and a high degree of integration, application, and management approaches.

To build the bulk goods service system requires the combination of specific content of the bulk commodity supply chain to consider the specific conditions of each industry chain, employ advanced technologies, and to incude the following subsystem functions: basic data management, a predictive analysis, a statistical analysis, operation analysis, and business intelligence. This connects with other business systems to share business data.

7.8.2 Business process and data flow analysis

1. Business process

This system extracts and transforms industry chain-related business data from the database, and combines modern management theories and optimization techniques to display business reports and analyze current business trends. It also evaluate the quality of business operations and forecasts business trends to support managers in decision making. This process is shown in Figure 7-23.

It extracts and transforms basic data in the industry chain, generates predictive models and manages historical data, performs statistics and analysis on the collected data, and applies data mining to perform multi-dimensional analysis, random inquiry, and display and business evaluation. Consequently, it provides decision support for the visualization of enterprise information, business information, and transportation routes.

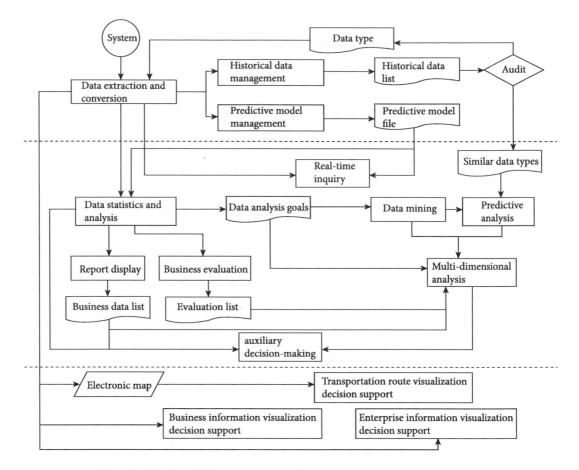

Figure 7-23   The business process of big data application service system

2. Data flow

According to analysis of the big data application service system and the classification of data in each industry chain, information available in each system is collected, extracted, integrated, shared, and processed to geneate the final data. This shown in Figure 7-24.

According to the supply chain management needs of logistics companies and the role of supply chain users, data sources are clarified from suppliers, manufacturers, sellers, logistics providers, and customers. And with preliminary analysis and processing by the data warehouse, the data process determined from warehouse supervision, comprehensive services, coordination control, and decision support.

7.8.3 Overall structure

According to a company's supply chain management business needs, data mining, IoT, and cloud computing are combined to extract, collect, integrate, and share data between different business systems, and design a big data application service system. The system includes subsystems for

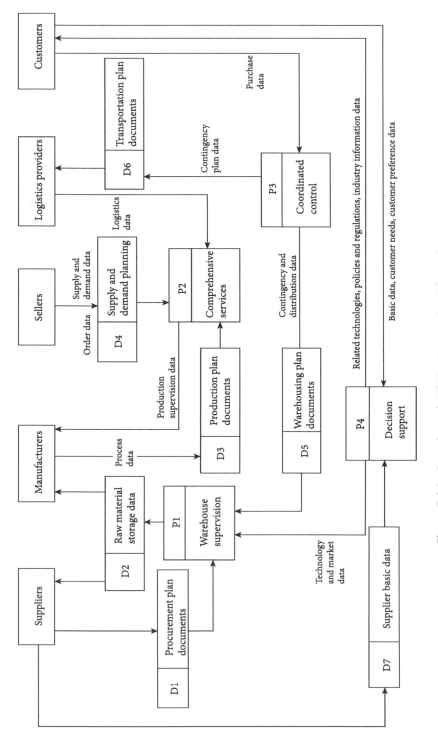

Figure 7-24　Data flow of the big data application service system

Note: P stands for related data processing departments, D for data information.

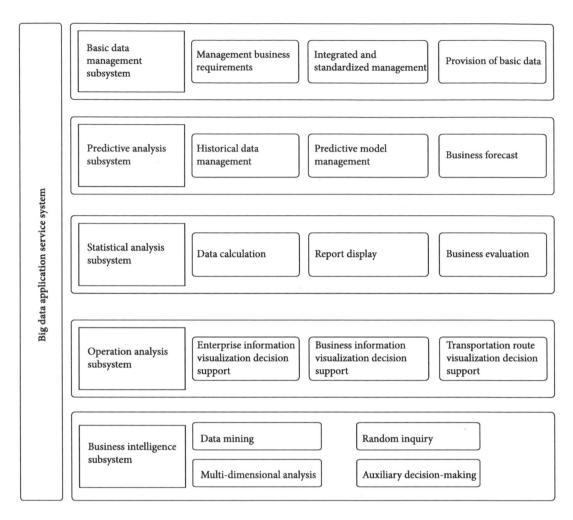

Figure 7-25   Overall structure of the big data application service system

basic data management, predictive analysis, statistical analysis, operation analysis and business intelligence, as shown in Figure 7-25.

The data obtained comes from the cooperative enterprise management, procurement service information, inventory service information, sales service information, supply chain financial service, and supply chain collaborative management information systems. The basic data management subsystem is responsible for obtaining and processing data of each enterprise in the industry chain; predictive analysis conducts business forecasting by managing historical data and forecasting models; statistical analysis is applied to the entire supply chain through data calculation, report display, and business evaluation; the operation analysis provides visual decision support for enterprise in the industrial chain; the business intelligence subsystem adopts advanced technology for data mining, random inquiry, multi-dimensional analysis, and auxiliary decision-making.

CHAPTER 8

BULK COMMODITY ELECTRONIC TRADING PLATFORM

8.1 Overview

8.1.1 Significance and objectives

The Bulk commodity trading platform refers to actual products with commodity attributes that are able to enter the circulation domain but not retail; they are traded in large quantities for industrial and agricultural production and consumption. In real trades, they have three types: energy and chemical products; minerals and metals; and agricultural and sideline products. The electronic trading of bulk commodities takes spot goods warehouse receipts (a valuable document of transaction) and synchronizes transactions of the same goods at different places through computer networks. The unified settlement is a form of market trading that organically combines the tangible market and the invisible market.

The development of the this bulk platform is based on Internet technology and e-commerce. It effectively connects bilateral customers, buyers and sellers, thus greatly reducing transaction costs and expenses, and improving convenience. As the inevitable product of bulk commodity market development, it matures as the corresponding trading market evolves. It has a high degree of information integration, a strong concentration of resources, and efficient and convenient trade circulation compared to fragmented trading markets. Consequently, it plays a greater role in economic development.

Building the platform is conducive to the rational allocation of resources and the establishment of an intensive development model. It helps stabilize the market and alleviate the operating pressure of enterprise imposed by irregular economic fluctuations, break industrial and regional price-fixing, promote marketization of commodity production and circulation; improve market circulation, and propel the development of the bulk commodity industry.

The trading, spot, and futures markets together constitute the Chinese modern commodity circulation system. They plays a vital role in China's fight for the pricing power of the international commodity market. In future, the Chinese bulk commodity market will become globalized, standardized, capitalized, large-scale, specialized, integrated, and intelligentized. Internet + bulk commodities will usher in a new era as a compelling chapter in Internet civilization.

8.1.2 Overall framework

The bulk commodity platform embraces seven subsystems: category management; e-commerce; member service management; market conditions; trade finance; corporate credit and risk management; and big data application service. It also deals with all sorts of enterprise related to bulk commodity transactions. Bulk commodity transaction services enable business interactions between enterprise and suppliers, sellers, and customers, who negotiate, trade and pay online. Thus, logistics and fund flow are of high-efficiency, information management is of low-cost, and operation is networked. The framework is shown in Figure 8-1.

8.2 Bulk commodity category management system

8.2.1 Demand analysis

Category management is a series of activities the bulk trading platform performs to divide commodities into different categories, each of which is treated as the "basic activity unit" of the platform's business strategy. For the platform, it improves operational efficiency and effectiveness; for platform users, it makes the selction of the bulk commodities they need more convenient and quicker while ensuring high quality. In addition, by managing commodity categories, the connections between logistics companies and upstream supply companies are consolidated.

Regarding implementation, category management sorts related products into one category for management by analyzing a large amount of data and formulating category strategies and tactics. The products' relevance is not determined by their inherent attributes (such as use, packaging volume, etc.), but the purchase behavior of consumers – especially which products they buy regularly during procurement.

8.2.2 Overall structure

The category management has five subsystems: basic information collection; category definition and identification; category catalog system building; category management; and statistical analysis, as shown in Figure 8-2.

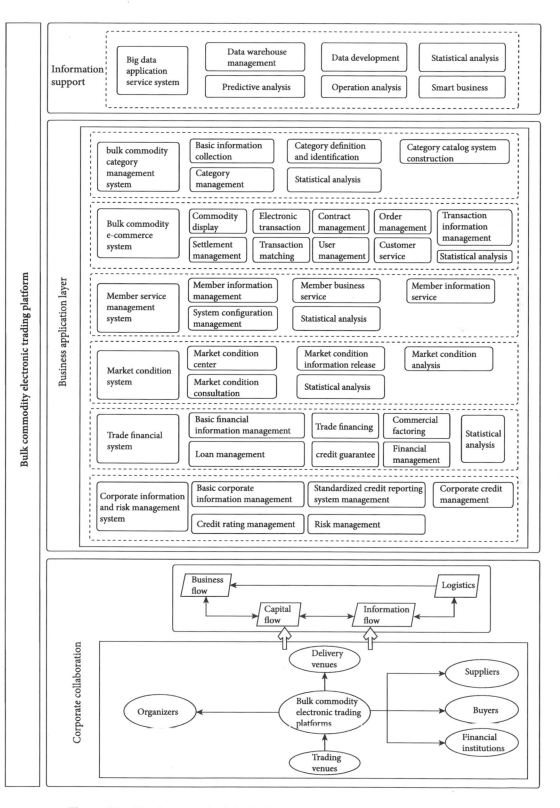

Figure 8-1 The framework of the bulk commodity electronic trading platform

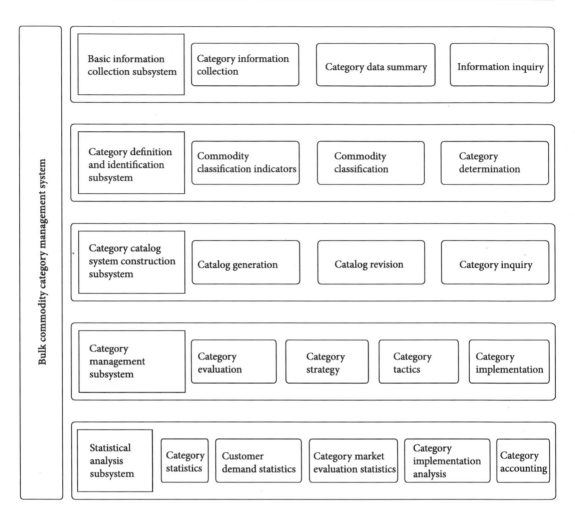

Figure 8-2 Overall structure of the bulk commodity category management system

1. Basic information collection

It collects, organizes, and summarizes various bulk commodity category information to provide data for catalog construction, including category information collection, category information summary, and information inquiry.

2. Category definition and identification

Category definition consists of category description and structure. Category description refers to classification standards by which the types of bulk commodities are classified – generally into energy and chemicals, minerals and metals, and agricultural and sideline products. Category structure refers to the proportional relationship between different types of commodities – especially the ratio between commodity types purchased by consumers. As for the criteria of classification indicators, product categories are located and divided, and subsequent category roles are determined and strategies implemented.

3. Category catalog system building

The commodity category catalog refers to the document that fully records and reflects the commodity classification system with tables, texts, numbers, and letters on the basis of commodity classification and coding. It is the primary basis for companies, according to their sales objectives, to formulate product purchase and sales plans by fixing the types of commodities that should be operated in a certain written form.

In addition to displaying the business scope of the company's products, it can also reflect the company's product level, principal categories, and sales profit sources. By systematically collecting, sorting, and classifying various description parameters of bulk commodities, a complete category catalogue is made, which is convenient for systematic category management. Also, products can be divided into latest products, best-selling products, or special products according to their transaction characteristics. This is convenient for online marketing and attractive to customers.

4. Category management

Organizing commodity accounting and management according to commodity categories is what modern commodity management requires. Category accounting is an innovation of the traditional commodity accounting system. Category management is a new type of product management that aims to meet consumption demand goals and connect supplier management. It also adopts modern information technology, is consumer demand-oriented, and aims to improve customer satisfaction.

5. Statistical analysis

Statistical analysis of customer needs, key product sales performance, category market evaluation, and other indicators constitute the basis for category management to improve.

8.3 Bulk commodity e-commerce system

8.3.1 Demand analysis

Since trading is the core platform business, the bulk commodity e-commerce system is one of its most core systems. Throughout the supply chain, where order flow acts as the principal line, it is a branch process. Generally, it is an online electronic trading platform that provides both buyers and sellers with online processing services in commodity transactions. For the transaction needs of upstream sales and downstream demand companies, it manages and controls all transaction links from contract generation to transaction settlement. It can help customers monitor transaction risk while providing them with opportunities for timely feedback on transaction results as a convenient electronic trading platform.

8.3.2 Overall structure

Based on the analysis of the bulk commodity business process and specific conditions of various bulk commodity trading companies, the bulk commodity electronic trading platform is designed. It includes ten subsystems: commodity display, electronic trading, contract management, order management, settlement management, transaction matching, transaction information management, user management, customer service, and statistical analysis. This is shown in Figure 8-3.

1. Commodity display

The bulk commodity e-commerce system builds a functional commodity display platform on the main page so that users can more easily and quickly understand commodity performance and characteristics. It therefore plays an advertising role. The transaction information display function includes sales and purchase information. The former enables buyers to purchase after browsing product sales information, while the latter allows sellers to sell after browsing purchase demand information. Information of new products needs to be added in the database punctually, while old data should be modified or deleted. Meanwhile, webpages should be updated and a product information database established to classify products and enable product searches to help buyers find what they need.

2. Electronic trading

With three functional modules: online transaction, transaction management, and transaction data inquiry, it is a good online trading venue for buyers and sellers.

Bulk commodity online trading, also known as online spot trading or spot hold trading, makes use of the same goods in different places, synchronous centralized bidding, unified matching, unified settlement, and real-time price quotation all organized by a computer network. It adopts a unique B2B business model because of the specific conditions of the Chinese spot market. As a mutually beneficial model that combines online and offline, real and virtual, and tradition and new economy, it fully solves bottleneck problems of information sources, customer sources, online settlement, and logistics.

Bulk commodity electronic trading management runs through the entire transaction process, which it controls and manages in accordance with relevant transaction management systems and regulations. A great deal of transaction data such as commodity price index, transaction volume, and trading area is involved in this. Among them, the price index is the paramount indicator, and its authority depends on the comprehensiveness of data collection. Setting transaction data inquiries on the platform helps users quickly understand the trading market and make better decisions.

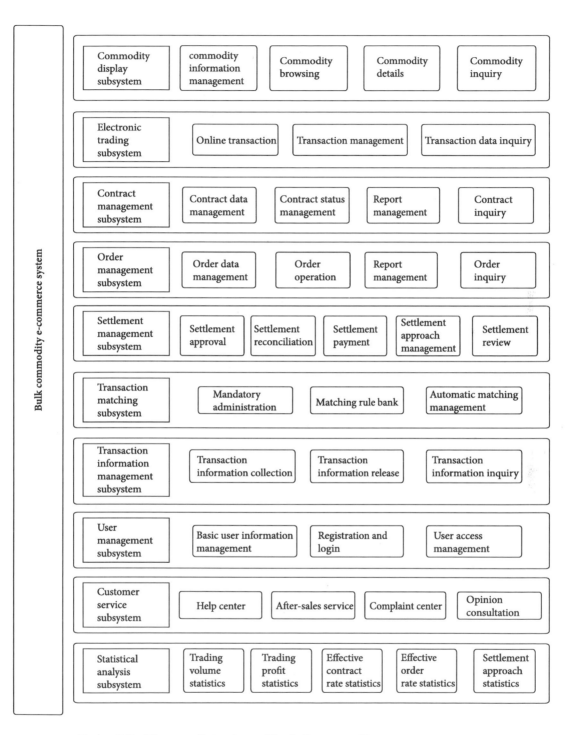

Figure 8-3 The overall structure of the bulk commodity e-commerce system

3. Contract management

Both buyers and sellers use the bulk commodity e-commerce system for online negotiations on all transaction details and authentication of the trade contract in the form of electronic documents. The negotiation results between the two parties include contract data management, contract status management, report management, and contract inquiry.

4. Order management

It includes order data management, order operation, report management, and order inquiry. Buyers can view the goods to be purchased in the shopping cart, and add, delete, and modify them. Once an order is confirmed, they can perform basic operations on the order, including inquiry, browse, modification, deletion, and generation of logs of related operations; in the environment set by the order administrator, the system can track the order status in real time, and feedback progress to the customers; customers can submit a return request where the system should provide them with return services and progress inquiries. According to the orders, sales, product sales law will be analysed as a basis for corporate decision-making.

5. Settlement management

A series of processes such as settlement approval, reconciliation, payment, method management, and review ensure the safe and rapid progress of transaction settlement. Settlement management can also deal with payment methods. When customers complete an online transaction, the electronic transaction business will provide them with post-transaction settlement services, which will send them detailed bill breakdowns.

6. Transaction matching

It is a means of exchange between multiple buyers and sellers. Buyers publish their purchase needs while sellers release information about products or services they can provide and their quotations into the system. Subsequently, an automatic matching procedure is activated according to certain transaction rules or the wishes of both parties. Once the matching succeeds, the transaction result is formed.

7. Transaction information management

Commodity transaction management utilizes computers to manage commodity transaction information, including commodities, transactions, procurement, and suppliers. Management maintenance of basic transaction information can add, modify, or delete various information of commodity transactions as well as search, collect, publish, and inquire.

8. User management

This includes business owners, customers, system administrators, etc. To ensure security, customers must log in to the system before purchasing products and placing orders. For the first purchase, they must first register and fill in relevant member information – this applies to business owners as well. The system will manage basic user information and set their access.

9. Customer service
It usually targets external customers of the logistics port with the purpose of rendering customers with better and more convenient services. They can obtain online help and after-sales service on this platform and file complaints about business owners if they are dissatisfied.

10. Statistical analysis
A statistical analysis is performed on data such as the transaction volume, transaction profit, effective contract rate, and effective order rate of each commodity type. Via a large amount of data analysis, customer consumption tendency and habits are understood. Next, the impact of transaction time, region, commodity price, and other factors are considered so that better personalized services can be provided.

8.4 Member service management system

8.4.1 Demand analysis

Membership management is a business model that acquires loyal customers and increases corporate profits in the long-term by rendering differentiated services and precise marketing. It will manage customer accounts, balance, points, member prepayments, member care, and data analysis. In fierce market competition, whoever grasps customer needs has the initiative. And the key to grasping customer needs is to subdivide members and launch precise marketing, so as to maintain good contact between business owners and customers, improv customer loyalty, and achieve performance growth. Meanwhile, it can also provide differentiated services for member users.

8.4.2 Overall structure

The member service management system strives to provide users with high-quality, personalized services that are different from ordinary users for mutual benefit. It consists of member information management, member business service, member information service, system configuration management, and statistical analysis, This is shown in Figure 8-4.

1. Member information management
It collects information and runs statistics on platform members. To become a platform member one has to register. It manages their access while allowing them to modify personal information.

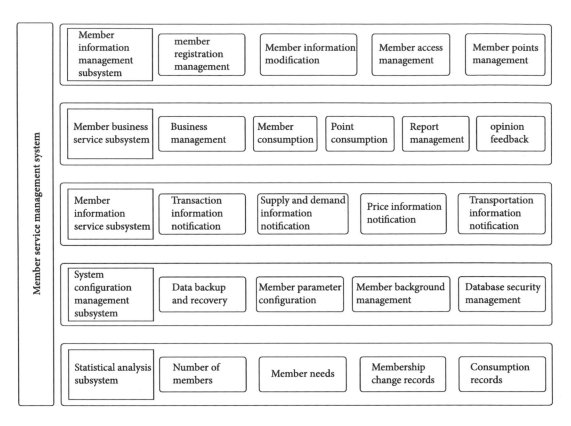

Figure 8-4   The overall structure of the member service management system

2. Member business service

Member business service consists of operational business management, member consumption management, point consumption, report management, and feedback.

Operational business management includes membership development and guiding members in system access.

Member consumption management mainly handles member consumption in different regions and network environments, and with different suppliers, shares extensive data, transmits member data, and collaborates. This improves overall work efficiency and responsiveness. It gives full play to management information by integrating this into the system, solves the problem of information sharing and full utilization, and provides primitive data support and decision analysis for platform management.

Point management means that when a member swipes a card that collects payment points, the system calculates the corresponding points that should be accumulated and writes them into the point account of the membership card. As points keep accumulating, members can enjoy certain preferential discounts when using the card. When the total points reach reach a certain level, the platform rewards them accordingly – in the forms of physical gifts and coupons.

Report management organizes member information in the form of reports, such as basic member information, consumption, and points, etc.

Feedback should be combined with database management and e-mail prompts so that information fed back can be processed in a timely and effective manner.

3. Member information service

It provide members with necessary commodity supply and demand information, transportation information, price quotation consultation, and product management, maintaining information symmetry between members and the platform. There are also additional modules of membership points management, member promotion, and friend management.

4. System configuration management and statistical analysis

System configuration management refers to data management that enables normal operation of the system. It includes data backup and recovery, member parameter configuration, member background management, and database security management. Statistical analysis is the use of statistical data and statistical methods, numbers and text for analysis. It embraces analysis of member numbers, needs, changes, and consumption records.

8.5 Market condition system

8.5.1 Demand analysis

Based on transaction data of all sorts of commodities in the present market, market and transaction information is displayed from multiple angles in a real-time manner, while internal connections between information sources are automatically established. This helps platform users to capture trading opportunities and provides them with expert opinions and online market consultation services.

A professional and unrestricted market analysis system should quickly, accurately, and fully display a variety of direct and mature technical charts, and provide customizable composition tools to meet the deeper needs of data warehousing and data mining. This requires information services to be immediate, complete, accurate, stable, and sustainable. At the same time, through deep data mining by the system itself, it can focus on risk presets, statistics, and alarms to intuitively express data change laws and provide an objective basis for users to make decisions.

8.5.2 Overall structure

Based on customers' different needs for market information, a market condition system is designed for the bulk commodity electronic trading platform, including five market sub-systems: a condition

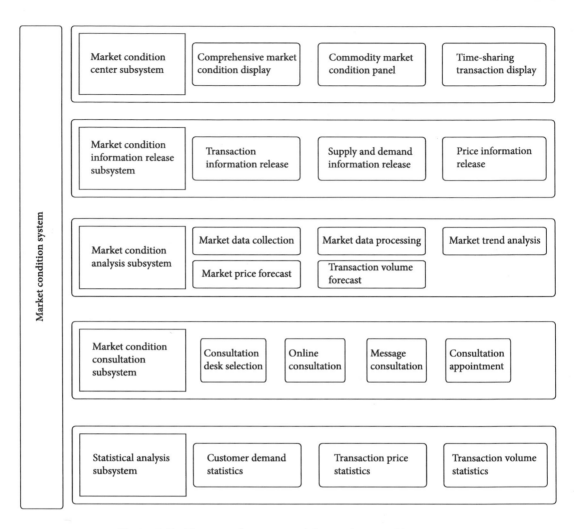

Figure 8-5   The overall structure of the market condition system

center, information release, condition analysis, condition consulting, and statistical analysis. This is shown in Figure 8-5.

1. Market condition center
It provides comprehensive industry information. With market condition displays, product market condition panels, and time-sharing transaction displays, it provides commodity transaction and cutting-edge industrial news fully known to platform users.

2. Market condition information release
The platform releases information on transactions, supply and demand, prices, etc. in a timely manner. Specifically, the information released includes commodity attributes such as product

type, origin, price, company, inventory, and remarks, etc. By exhibiting transaction information, customers can easily understand products and find matching choices.

3. Market condition analysis

By collecting, sorting, and analyzing market data, visual market trend conclusions are drawn, and key transaction data is therefore predicted. Generally, the distribution of transaction volume is presented in the form of diagrams and maps and the price index of the products in the form of dynamic tables, trend charts, etc. Therefore, customers know real-time prices and their changing trends, which facilitates their purchase behavior.

4. Market condition consultation

Users can communicate with and consult platform staff online through mobile messaging, direct dial, conference, etc. Big data analysis is adopted to support and integrate enterprises' internal expert resources to provide customers with professional assistance, including business consulting, plan design, demand analysis and forecast reports in the logistics industry.

5. Statistical analysis

Statistical analysis involves customer demand, transaction price, and transaction volume statistics. The data collected can be used for market condition analysis and evaluation, which is conducive to platform management and operation. To some extent, it can also provide users with bulk commodity trading market conditions and help them make better decisions.

8.6 Trade finance system

8.6.1 Demand analysis

The bulk commodity trade finance system revolves around principal financial products of the platform including trade financing, commercial factoring, loan management, and credit guarantees. This give strong financial and monetary support to platform users through digitalization; meanwhile, it can improve financing, factoring, and loan quality, leading to a multi-win outcome.

The trading finance system not only helps the platform and financial institutions to effectively monitor logistics and fund flows, and control risks, but enables spot companies to improve their creditworthiness through daily trade on the platform. At present, some large-scale bulk commodity spot trading platforms in China can effectively manage the purchase, warehousing, logistics, and sales of bulk commodities, establising a relatively functional supply chain system. This can help enterprise reduce operating and financing costs, thereby fundamentally improving its ability to serve the substantial economy.

8.6.2 Overall structure

This system is an information system designed and developed according to the financial business needs of logistics companies as well as the needs of real-time supervision of financial institutions. This improves capital operational efficiency, trade business, and financial business. Through systematic management of financial information and services, it improves the financial service quality and achieves mutual benefit for all. It is composed of basic financial information management, trade financing, commercial factoring, loan management, credit guarantees, financial management, and statistical analysis, as shown in Figure 8-6.

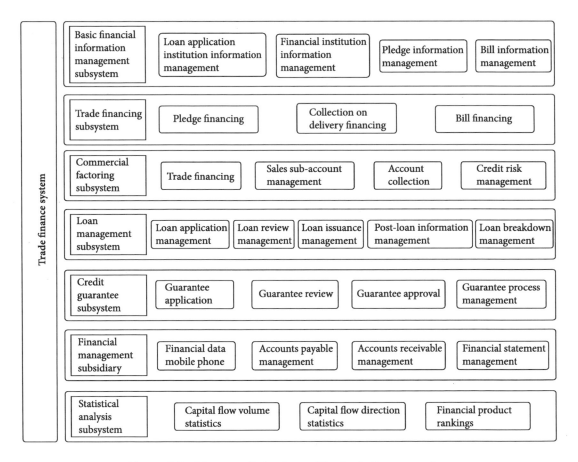

Figure 8-6   The overall structure of the trade finance system

1. Basic financial information management

It involves the information of multiple business participants, including loan applicants of financial institutions, pledges and bills, etc. The collection and management of basic financial information makes it easier to support the financial business.

2. Trade Financing

Generally, it is defined in two senses. Broadly, as long as the recipient of financing services is an enterprise engaged in bulk commodity trade, their business falls into the bulk commodity trade financing category. The narrow sense of trade financing refers to bulk commodity structured trade financing, that is, banks offer bulk commodity traders or production companies a personalized trade financing portfolioto for liquidity management and risk mitigation. It has three parts: pledge financing, collection on delivery financing and bill financing. These are financial support services for platform users: trade financing supports Small and medium enterprises (SME) and solves insufficient credit issues, which leads to poor financing channels and excessive monetary pressure. In addition, it can revitalize the capital occupation of temporarily idle raw materials and finished products in a company, optimizing its resource utilization.

3. Commercial factoring

It means the seller transfers the current or future accounts receivable based on the sales contract with the buyer to the factor, who renders a series of comprehensive services, such as trade financing, payment collection, and credit risk management. With systematic management, bulk commodity trade finance services make the factoring business more efficient, which is conducive to reducing bad debts and improving quality.

4. Loan Management

It includes management of loan application, loan review, loan issuance, post-loan information, and loan detail. The effective management and control of the loan business process can contribute to the efficient operation of each link.

5. Credit guarantee

It is a way of financial support that guarantees bank claims when companies finance funds from banks through a legally established guarantee agency, who vouches for the debtor in accordance with the contract. If one cannot service the debt in accordance with the contract, the guarantee agency shall bear the contractual responsibility for repayment. It includes guarantee application, review, approval, and process management. Throughout the guarantee business, management and control of each link should be well performed so that the credit guarantee service is rendered to those platform users in need.

6. Financial management

It systematically manages the financial transactions of the entire system by collecting financial data and managing account receivables and payables. It prepares financial statements and then clarifies each financial transaction, enabling the system to operate transparently.

7. Statistical analysis

This is run on the volume and flow direction of capital involved in financial products to ensure that when the platform conducts trade and finance activities, the fund flow and trade exchange between enterprise is clear, transparent and efficient. Screening big data can provide companies with the most suitable financial product information and financial trade services.

8.7 Corporate credit and risk management system

8.7.1 Demand analysis

Corporate credit risk refers to the possibility that one transaction party fails to pay as promised and causes a loss to the other transaction part when the credit relationship acts as their link. It's usually manifested when corporate customers do not pay for the goods or are unable to. Credit risk management refers to the implementation of comprehensive supervision and control of customer credit investigation, payment method selection, credit limit determination, and payment recovery by formulating credit policies. This guides and coordinates each institution's business activities in order to safeguard the timely recovery of receivables.

As the party that develops, operates and manages the platform, it is necessary to conduct corporate credit investigation and assessment on corporate users, create a credit management system in line with the market credit system, and create a unified standard system and regulatory basis to protect the legitimate rights and interests of all platform users.

8.7.2 Overall structure

The corporate credit and risk management system is composed of five management subsystems: corporate basic information, standardized credit reporting, corporate credit reporting, credit rating, and risk. This is shown in Figure 8-7.

1. Basic corporate information management
Basic corporate information management is performed for corporate platform users as the basic data for corporate credit reporting and credit rating. This includes basic information input related to user credit reporting, basic data management, and information maintenance and inquiry.

2. Standardized credit reporting system management
It concentrates on the characteristics of relevant company participants in bulk commodity transactions, and combines core indicators of other related corporates' credit reporting to determine the essential elements and principal indicators. Often, it includes credit reporting elements and standards, index weight determination, credit reporting methods, and credit review.

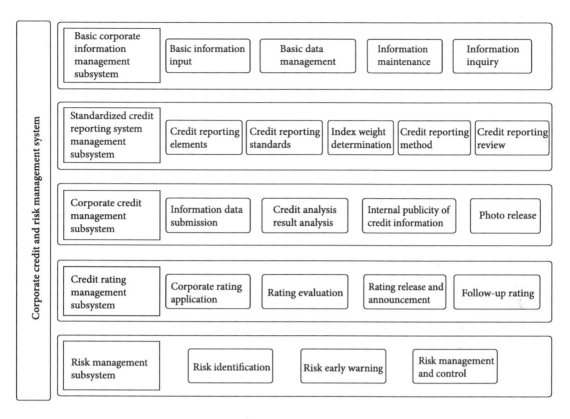

Figure 8-7 The overall structure of the corporate credit and risk management system

3. Corporate credit management

As the party that operates and manages the platform, it implements credit management for platform corporate users. By reviewing and evaluating the information and data, logistics companies release a list of corporates with reasonable credit and issues their license. It strives to help them maximize sales while minimizing credit risk, thus increasing their interests and value as much as possible.

4. Credit rating management

It is responsible for handling credit rating applications, evaluating rating, releasing rating, and follow-up rating. Objectively and systematically, it conducts corporate credit rating for other platform users to refer to. The corporate credit rating in China usually covers corporate qualification, capital credit, operation management, economic benefits, and development prospects.

5. Risk management

It is a method for companies to try and control the consequences of uncertain factors within an acceptable range in the process of achieving future strategic goals, so as to ensure and promote the organization's. overall interests. Risk identification, risk early warning, and risk control can prepare them better to reduce losses.

8.8 Big data application service system

8.8.1 Demand analysis

The big data application service system adopts big data technology to analyze and mine the data in detail, thus supporting and safeguarding platform system operation. Big data will be used from order generation to transaction completion and to the final result feedback. It can fully control the entire transaction and allows a clearer grasp of the demand, inventory, order completion rate, and transaction satisfaction of both parties; meanwhile, it can also provide auxiliary decision support for platform users.

The bulk commodity trading market is at the center of the commodity spot market. It directly faces industrial customers. At the forefront of production and trade, it assume the dual roles of a platform for market parties to conduct price games and commodity trading activities, and an important link with tremendous advantages in data collection for entities to connect with the financial industry. As an essential part of the national big data strategy, it has access to both data resources of the entire industry and rich financial data. In addition, market participants also have demands for information and data. Therefore, competent data mining and analysis, stronger risk control, business innovation and customer service capabilities are urgently required in the bulk commodity trading market.

8.8.2 Overall structure

This system consists of six subsystems: data warehouse management, data development, statistical analysis, predictive analysis, operational analysis, and smart business. This is shown in Figure 8-8.

1. Data warehouse management
Data warehouse is a structured data environment for decision support systems and online analysis application data sources. Subject-oriented, integrated, stable and time-varying, it studies and solves the issue of obtaining information from the database. It is able to clean, extract, convert, and organize a lot of data in the database, which is used for transaction processing according to the needs of the decision-making entity, and meets the requirements of diversity analysis. It includes product information, transaction, and customer data. To manage the data warehouse involves data collection, data display, online analysis and log collection.

2. Data development
Collected data is imported into the background platform system for development. Next, an in-depth study of core database related technologies is conducted to design and build a database management system. When the business requirements of database applications are understood, the existing database architecture keeps being optimized. Database logic and physical models are

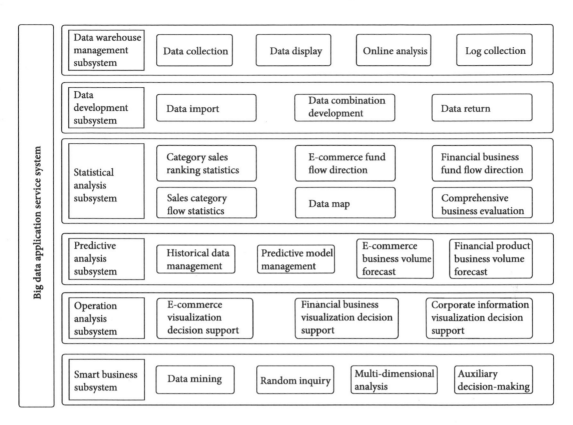

Figure 8-8   The overall structure of the big data application service system

therefore designed according to business requirements, and storage procedures, functions, and scripts required by database production environment are developed.

3. Statistical analysis
Statistical analysis involves many types of data, such as category sales ranking statistics, e-commerce, financial, and sales statistics, which are sorted and analyzed to make a data map.

4 Predictive analysis
According to business data, modern forecasting methods and technologies are combined to design and construct a predictive analysis subsystem, whose functions are historical data management, predictive model management, e-commerce business volume forecasting, and financial product business volume forecasting.

5. Operation analysis
Based on the basic information of enterprise, commodities and transaction, GIS technology is applied to design and build an operation analysis subsystem, which is able to provide visualized decision support for e-commerce, financial business, and enterprise information.

1) E-commerce visualized decision support
It adopts electronic maps or graphical visualization to display supply and demand information of platform customers, as well as purchase and sales information, and to manage related transaction information.

2) Financial business visualized decision support
It adopts platform or graphical visualization to display the supply and demand information of financial services and financial information of related logistics.

3) Enterprise information visualization decision support
It adopts platform or electronic map visualization to display supply and demand information of company participants in transactions and their related business process information.

6. Smart business
Relevant statistics, analysis and inquiry results are combined, through automated indicators and report information management, to perform a comprehensive and detailed analysis of business operation quality, business risks and management levels, and generate smart decision-making data. Therefore, a comprehensive range of multi-level decision-making support is made available to business managers.

1) Data mining
With relevant algorithms and tools, the hidden, unknown and potential information or data patterns are extracted from the database; business data is transformed into business information with reference and application value; and smart decision-making support for related businesses is made available.

2) Instant inquiry
According to user instructions, real-time inquiry is made about model information, parameter information, indicator information, and assessment results of key performance indicator (KPI) management to provide relevant personnel with prompt and accurate decision-making support.

3) Multi-dimensional analysis
Business intelligence views, perceptual information views, smart decision-making results, and user inquiry models are collected, analyzed, and displayed in a multi-dimensional form to enhance business data readability and provide full and accurate information support.

4) Decision-making support
Relevant statistics, analysis, and inquiry results are combined, through automated index and report information management, to provide comprehensive and multi-level decision-making support and knowledge services. This assists relevant personnel in detailed analysis of business operation quality, business risks, and management levels.

CHAPTER 9

LOGISTICS PARK INFORMATION SERVICE PLATFORM

9.1 Overview

9.1.1 Present situation of logistics parks in China

As the logistics industry blooms, logistics parks have attracted incremental attention from all sectors of society regarding their development and construction. In 2009, the State Council issued the Adjustment and Revitalization Plan of the Logistics Industry; in 2013, the National Development and Reform Commission and releted governmental departments jointly issued the National Logistics Park Development Plan (2013–2020), clarifying their objectives, overall layout, and primary tasks; in 2014, the State Council issued the Medium and Long-term Plan for the Development of Logistics Industry (2014–2020), proposing that by 2020 a modern logistics service system with reasonable layout and advanced technology, convenient and efficient, green, safe, and orderly will have been established. As governments at all levels have regarded the construction of logistics parks and logistics industry advancement as key measures for regional economic development, logistics parks are being planned and constructed one after another.

According to the Fourth National Survey of Logistics Parks (Bases) issued by the China Federation of Logistics and Procurement and the Chinese Society of Logistics in July 2015, the country had a total of 1,210 logistics parks in operation, under construction, and planning. Compared with 207 in 2006, it experienced an increase of 484%; with 475 in 2008, an increase of 155%; with 754 in 2012, an increase of 60%. The number is shown in Figure 9-1.

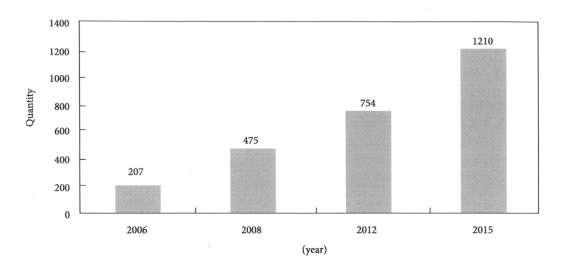

Figure 9-1   The number of logistics parks in China

The results of the survey report show that parks with an informatization investment of less than RMB 10 million accounted for 72%, and those with less than RMB 1 million made up 34%. The total investment is shown in Figure 9-2. Although informatization is still developing quickly, there is still a lot of room for improvement regarding informatization.

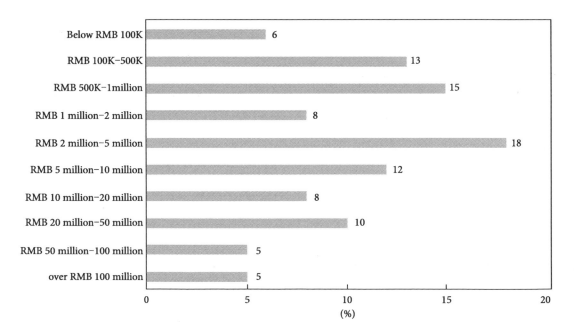

Figure 9-2   Total investment in logistics park informatization

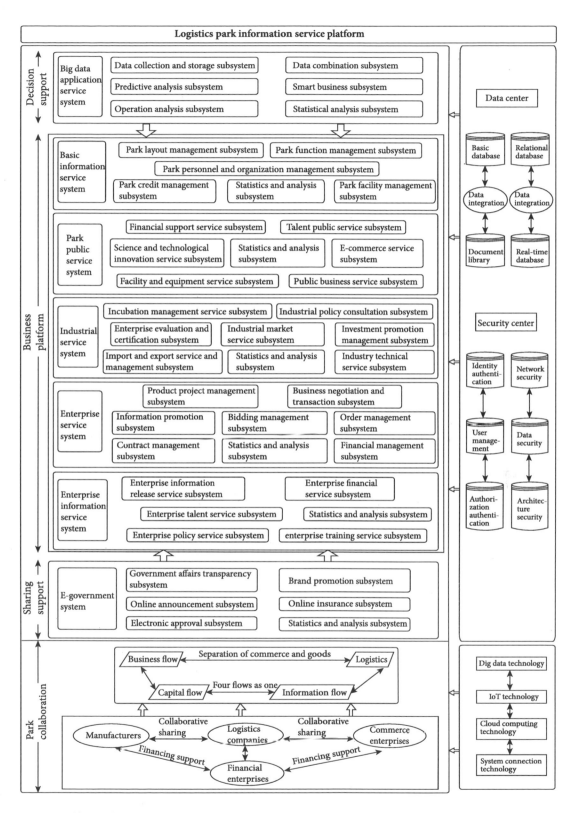

Figure 9-3   The overall structure of the logistics park information service platform

9.1.2 Introduction

The logistics park information service platform is divided into 7 major sections: basic information service, enterprise information service, enterprise business, industrial service, E-government, public servic, and big data application service. Involving 50 subsystems, it enables the seamless connection between local government, logistics parks, and enterprises, thus effectively promoting information sharing and communication and improving problem solving. It aims to build a resource sharing platform for the park management committee and the settled enterprises to integrate, manage, and share information resources within the park. This demonstrates advanced technology and informatized management to serve industrial aggregation, resource sharing, and long-term development. This is shown in Figure 9-3.

9.2 Basic information service system

9.2.1 Demand analysis

1. Background and significance
Basic information management and services are the platform's foundation to realize system functions. By managing the basic information of logistics parks, information is connected and shared within them; their informatization is accelerated and an interconnected information resource network is built. To promote industrialization with informatization is an important part of propelling the logistic park development. This also serves sustainable development. The basic information service provides necessary information for the park management committee, settled enterprises, and related individuals. Users at different levels can inquire about relevant information through the platform.

2. Functional requirements
The basic information service system is the data core of the comprehensive information service platform. It is responsible for the unified management of basic park information as well as the provision of information and data inquiry services to external users. With a complete data set, analysis and processing, it meets the needs of logistics companies and relevant government departments for basic information services, and provides corresponding information services as required by different users.

9.2.2 Overall structure

Based on the principles of openness and resource sharing, management services including park layout, function, personnel, and organization, credit, and facilities are provided as well as statistical analysis to management committees, enterprise, organizations, and individuals in the park. This is shown in Figure 9-4.

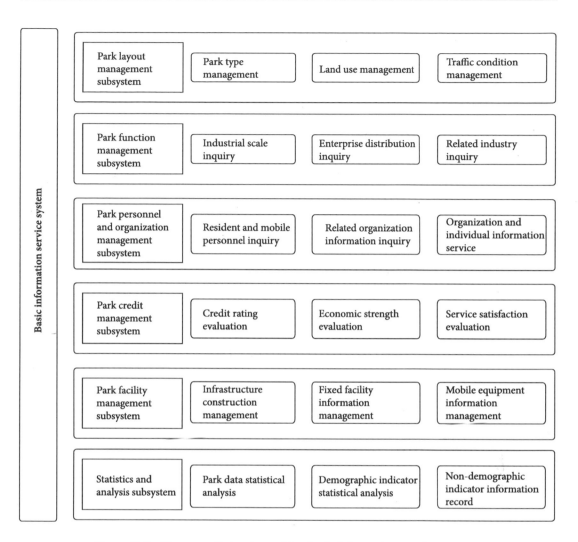

Figure 9-4 The overall structure of the basic information service system

1. Park layout management subsystem

It records the overall layout and industrial scale of the logistics parks and can launch enquiries on basic parameters such as park type, land use, building scale, road layout, supporting facilities, etc. Based on their layout requirements, standardized, normalized, and centralized management approches are combined to establish the park layout management subsystem. This ensures smooth operation of various businesses in the logistics parks.

2. Park function management subsystem

It records the functions and capabilities of the logistics parks. It can inquire about basic conditions of their specific functions – industrial agglomeration, spatial pattern, enterprise distribution, and industrial associations. Based on functional development needs, the general management methods of their functions are combined to build the park function management subsystem. The preparation of function management should integrate the actual requirements of the management committee, related enterprise, and individuals.

3. Park personnel and organization management subsystem

It records logistics park personnel and institutions, and can inquire about personnel-related information such as flow, positions, and credits. This makes it easier for enterprises and individuals to contact relevant departments and institutions, thereby ensuring the functional activities of the organization's goals and tasks; it is required to accurately record, reasonably frame, and manage related organizations and personnel, including resident and mobile personnel inquiry, related institution information inquiry, and institutional and individual information services.

4. Park credit management subsystem

It records the credit evaluation of the logistics parks and can inquire about service evaluation, rating evaluation, economic strength, development prospects, and other information related to their credit level. To reduce park management loopholes and park and enterprise risks, it collects and manages basic credit information in multiple ways and evaluates park credit ratings, economic strength, and service satisfaction evaluations.

5. Park facility management subsystem

It records the facilities and equipment of logistics parks so that relevant entities can grasp the situation of physical resources, property, and facilities, including information on infrastructure, fixed facility, and mobile equipment. There are a great variety of infrastructure and fixed facilities in the park, therefore, to ensure safe business operation, this subsystem is established. It includes infrastructure, fixed facility information, and mobile equipment information management. It can inquire, add, modify, and delete information.

6. Statistics and analysis subsystem

It collects and organizes basic information service activities data, analyzes it based on park development needs and the status of basic information services, and feeds back the results to government managers and related enterprise in the form of reports and documents. This helps users analyze park services and understand the needs of related enterprise and individuals, thus improving basic information management in the parks.

9.3 Enterprise information service system

9.3.1 Demand analysis

1. Background and significance
The enterprise information service system is an important platform that guides management of relevant entities and serves enterprise in the logistics park. According to the principles of governmental guidance, park management, and market operation, advanced concepts and technologies are adopted to record and promote the park and settled enterprise on the information platform; efforts are gathered to create an information service platform with unique influence and functions. It meets the various needs of logistics parks for policy guidance, work deployment, and coordination of economic operation elements. Ultimately, a fast and convenient communication channel is bridged between the logistics park management committee and enterprise.

2. Functional requirements
This system has six components: talent service; financial service; policy service; training service; statistical analysis, and related information release. (Each of these has three corresponding functional modules.) Management committees and enterprises in the park can inquire about information services on the platform. These information services are divided into categories: those related to basic, static, and dynamic enterprise information and those related to talent, finance, policy, and training.

9.3.2 Overall structure

This the most comprehensive management information system in the park. Supported by the park's service system and interconnected with various information systems such as government departments and financial service institutions. It enables information exchange between park enterprise and provides information release, talent, financial, policy, and training services, and statistical and analysis. This is shown in Figure 9-5.

1. Enterprise information release service subsystem
It releases information to the park and can make inquires about business information, advertising, press releases, and recruitment information. Based on their business development needs, the general approaches of information release management are combined to establish an enterprise information release service system; this ensures their normal operation. Information release must integrate and screen all types of information from each enterprise, including basic, static, and dynamic information release.

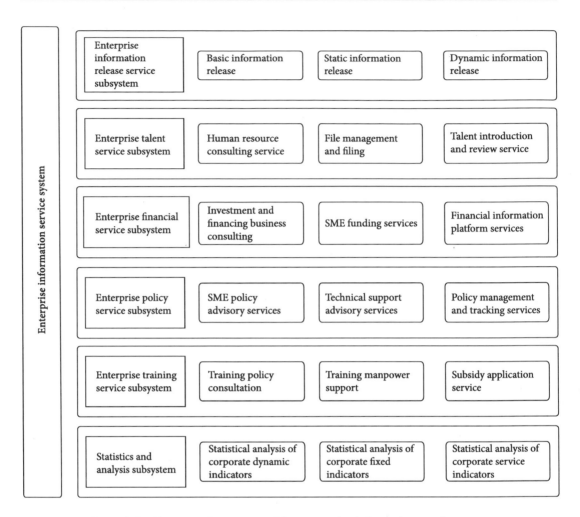

Figure 9-5 The overall structure of the enterprise information service system

2. Enterprise policy service subsystem

It renders national and local policy consulting services for parks and handles the application acceptance, tracking management and services of supportive policies, such as, technological development funds and leader plans. To ensure that all logistics park enterprise enjoy the support of national and local policies promptly, a comprehensive system is established. This includes policy advisory service for small and medium-sized enterprise, technical support advisory services, and policy management and tracking services to perform policy consultation, application acceptance, follow-up management, etc.

3. Enterprise talent service subsystem

It provides a one-stop service platform for enterprise talent management, including human resource market inquiry, file management and archiving, and talent introduction and review. Based on their need for talent flow, the interconnection of external talent market information helps greatly. Specifically, it covers human resources consulting, file management and archiving, talent introduction and review, etc.

4. Enterprise financial service subsystem

The financial service industry, as a financial product trading platform, analysis platform, and investment and wealth management channel, provides users with professional and real-time financial information. It uses this to create greater value in financial activities and solve financing problems in the parks. This subsystem effectively injects the convenience of the Internet into financial information services. The subsystem provides various financial management services: investment and financing business consulting; financial support for small and medium-sized enterprise; and policy and information support for enterprise engaged in financial business.

5. Enterprise training service subsystem

It provides enterprise with a one-stop service for employee training, including policy consultation regarding rare talent, project idea collection, subsidy applications, etc. Based on their need for talent, modern methods of employee training management are combined to establish a enterprise training service system, including training policy consultation, training manpower support, and subsidy application service.

6. Statistics and analysis subsystem

By collecting and sorting out the service activity information of logistics park enterprise and analyzing the data as per the development goals, statistical analysis services of dynamic, fixed, and service indicators are rendered to help managers. This helps understanding of development strategy goals and business planning, diagnosis, and evaluation of enterprise management status, optimizes business process, and proposes visions, goals and strategies for enterprise growth.

9.4 Enterprise business system

9.4.1 Demand analysis

1. Background and significance

In the present development of informatization in China, the status quo of enterprise informatization construction in logistics parks is far from ideal, with many problems in urgent need for solutions. As requirements vary in different industries, there are many problems to consider in the construction of logistics parks, which makes it difficult to form a unified business management model and greatly hinders interconnection. Meanwhile, due to the lack of corresponding standard guidelines, logistics standardization in China has been seriously delayed, which leads to very little application of information technology in logistics. As a result, logistics information is scattered, resources are ineffectively integrated, and information is isolated, which impedes further logistics park informatization.

2. Functional requirements

It has eight subsystems: product project management, business negotiation and transaction, information promotion, bidding management, order management, contract management, financial management, and statistics and analysis. All types of logistics park enterprise can handle business between themselves through this platform and can cooperate. It allows them to use the Internet to build an image, promote products and business operations, and set up an internal information management platform.

9.4.2 Overall structure

The enterprise business system developed for logistics parks and enterprise is a set of simple and practical office automation solutions based on fixed and mobile networks. It is designed to help parks and enterprise customers to efficiently build an integrated collaborative office platform via the Internet and wireless network, and a convenient and fast communication channel. It enables key enterprise to report important data, information, and major emergencies of the logistics parks, and creates a collaborative business circle between parks and enterprise. The overall structure is shown in Figure 9-6.

1. Product project management subsystem

It approves, distributes, inquires and manages enterprise product projects. Specifically, administrators define or change and distribute projects according to the situation and approval procedures; to ensure the standardization of engineering projects, throughout the project management module the subsystem provides inquiry methods for project management and the number of documents that users count. From the beginning to the end, the entire project is planned, organized, commanded, coordinated, controlled, and evaluated to achieve its goals.

2. Business negotiation and transaction subsystem

As a negotiation system, it enables business negotiation, information creation, sending, receiving, forwarding, reading and replying, and uploading of all types of attachments. Meanwhile, as a transaction platform, it renders trading and filing services for enterprise and enables them to cooperate and trade. There are mobile office systems based on WAP, Linux, and Window that launch automatic upgrades of client programs. It include business negotiation management, business transactions management, and business filing services.

3. Information promotion subsystem

It customizes an information promotion platform for enterprise in the logistics park to promote information related to their business management, and improve information promotion efficiency

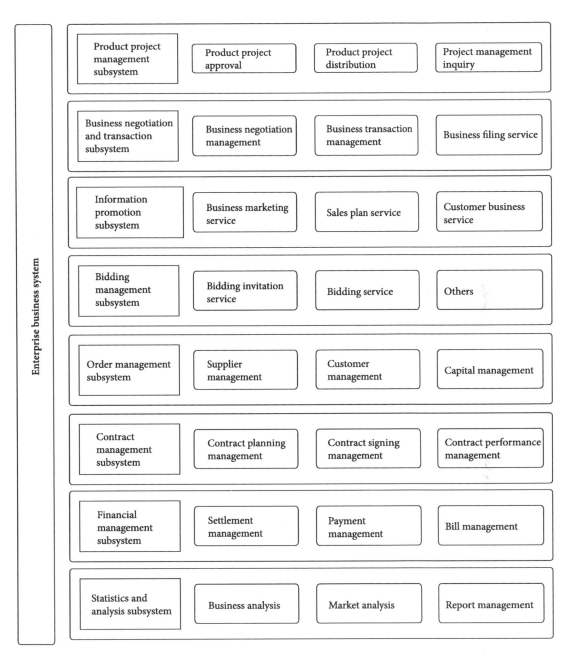

Figure 9-6 The overall structure of the enterprise business system

between them. Specifically, there are business marketing, sales plan, and customer business services. By managing customer files, sales leads, sales activities, business reports, and statistical sales performance, it assists sales managers and staff in quickly processing important data.

4. Bidding management subsystem

Based on the latest Internet technology, it adopts the current advanced B/S structure to introduce web technology into bidding management of a project. It aims to be open, transparent, fair, just, reasonable, honest, and credible. Accessible to all enterprises, it serves their project bidding by providing an information operation platform and improves project cooperation efficiency. The bidding management system includes an invitation for bids service.

5. Order management subsystem

It manages and tracks orders placed by customers, dynamically grasps their progress and completion, and improves logistics efficiency, thereby saving operation time and costs and raising the market competitiveness. Through unified orders, it provides users with integrated one-stop supply chain services. Order management and order tracking management suffice for users' logistics services. It performs regular verification and quantitative management of business between enterprise, provides a convenient operation platform for business cooperation, and improves their efficiency. By management objects, it can be divided into supplier, customer, and capital management.

6. Contract management subsystem

With information technology, it adopts advanced management ideas of modern enterprise to create a comprehensive and systematic contract management platform for decision-making, planning, control, and business performance evaluation. It manages and transmits contracts between enterprise to ensure their normal business cooperation and specifications of legal transactions. It involves contract planning, contract signing, and contract performance management.

7. Financial management subsystem

In addition to accounting service and financial operation provisions in traditional financial management systems, such as general ledger management and financial statement preparation, it also operates modern financial management – individual income tax calculators, fiscal budgets, etc. This manages the financial affairs within an enterprise and between enterprises to ensure standardization and unification of accounts. Established enterprise networks are utilized to disclose their financial system, budgeting, and policy basis, items, standards and scope of charge. It can, on behalf of companies and customers, inquires about income and expenditure and pay employee wages, allowances, and taxes.

9.5 Industrial service system

9.5.1 Demand analysis

1. Background and significance

The industrial service system is one of the core factors affecting industrial agglomeration, while the industrial service platform is an essential component of the industrial service system. It revolves around leading enterprise in the logsitics park to offer supporting services, including R&D, testing, equipment sharing, inspection, entrepreneurial incubation, and investment financing related to production, operation, and management and sales.

At present, the supporting service system is incomplete, which hinders deeper development of industrial chains and makes it difficult for enterprise to grow bigger and stronger. With low industrial relevance, short industrial chains, and difficulty in generating a cluster effect, it cannot work its magic as it should. However, the construction of the industrial service system can solve these problems, extending industrial chains and promoting industrial agglomeration.

2. Functional requirements

The system embraces seven subsystems related to industries in the logistics park: incubation management service; industrial policy consulting; enterprise evaluation and certification; industrial market service; investment promotion management; import and export service and management; industrial technical services; and statistics and analysis. Industry-related institutions or departments can serve leading industries there.

9.5.2 Overall structure

Based on the overall development strategy and guided by complex functions, an industrial service system is created, taking industries within the park as the foundation and combining industrial service functions to integrate industrial services. This renders logistics services that park industries, determine the main operation path, gradually drive regional development of core economic zones, and finally realize overall development. This is shown in Figure 9-7.

1. Incubation management service subsystem

The business incubator is a new type of social economic organization. It offers venues to R&D, production, and business operation; shared facilities to communications, networks and offices; and support to system training and consulting, policy, financing, legal affairs and marketing. As a cradle for cultivating SMEs and entrepreneurs, it lowers the entrepreneurial risks and costs of start-ups and improves their survival and success rate.

The subsystem, as a major carrier to promote the transformation and industrialization of scientific and technological achievements, provides comprehensive industrial incubation management and one-stop incubation services for the logistics park as well as incubation services

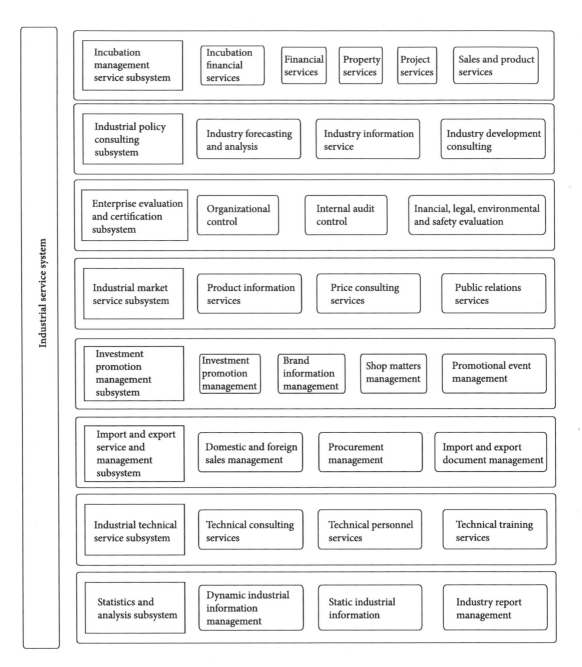

Figure 9-7 The overall structure of the industrial service system

for incubation projects, enterprises and individuals. It consists of incubation financial, property, project, sales and product services, etc.

2. Industrial policy consultation subsystem

Industiral policy consultation refers to the research activities that draw policy-related optimization plans or support decision-making of enterprise in the park by studying and analyzing strategic, comprehensive, and long-term national and regional policies. That is to say, they perform statistical investigation and research for enterprise to provide systematic background information, analyze and estimate the situation, forecast development trends, and propose strategic and political propositions and measures.

The subsystem forecasts, analyzes, and advises park industrial development, including the provision of industrial development information for park managers, industrial analysis information and industry reports, and information regarding industrial channels, business layout, industrial macro dynamics, and industrial planning.

3. Enterprise evaluation and certification subsystem

It renders credit evaluation and certification services for all types of enterprise in the logistics park. It creates an independent credit rating agency that has been rigorously reviewed and approved by regulators. With standardized evaluation and certification procedures, it independently, objectively. and fairly evaluates companies' credit rating to help the park master credit information. It has the following functional modules: organizational control, internal audit control, financial evaluation, legal evaluation, environmental evaluation, and safety evaluation.

4. Industrial market service subsystem

It studies the industrial market of the logistics park. Based on the development characteristics of major industries, it aims to build a complete and superior industrial chain. Considering the actual situation of companies in the park, led by the government, it takes enterprise as the main operating body and adopts market-oriented operation means to plan and aggregate while providing resource complexes, including capital, technical, talent exchange, and market trading platforms. This drives the development of enterprise to push forward industrial upgrade.

Specifically, it provides many types of market consulting and service for park industries, including product information, price consulting, industry supply and sales information, regional market analysis, customer consulting, and public relations. These timely and convenient services will help them open up the market.

5. Investment promotion management subsystem

A market-oriented investment promotion mechanism is the most effective way to rapidly and scientifically develop economy of the logisitcs parks. External forces are used to attract business and explore investment promotion potential; cooperation is bridged with professional investment promotion agencies; market-oriented means are adopted to optimize resource allocation and capital efficiency; in terms of project docking and talent introduction, all-round cooperation is conducted among enterprise in the region.

The subsystem provides informatization conditions and technology for investment promotion of the park, including investment, brand information, store, and promotion management, and renders policy, land, registration, investment, and taxation policies to related parties.

6. Import and export service and management subsystem

This refers to ERP (Enterprise resource planning) and SCM (supply chain management), which related enterprise to import and export perform electronic and functional management for circulation of funds, information, and commerce within their business scope.

It provides import and export business management and services for industries and related enterprise in the logistics park and greatly improves the level of import and export informatization. There are functional modules such as domestic and foreign sales management, procurement management, and import and export document management.

7. Industrial technical service subsystem

It is a support system that renders technical services for industrial development. By integrating separated technical resources in government and enterprise, it enables the sharing of industrial software and hardware equipment, technical information resources, and science and technological talent. Consequently, it reduces the risk of technological innovation of enterprise in the park, thereby lowering operating costs and accelerating industrial development.

It covers technical consulting and services for industries in the park, including technical requirements, technical consulting, technical talent services, industrial technical standards, information services, technical training, technical documentation services, and platform management.

9.6 E-government system

9.6.1 Demand analysis

It enables the logistics park management committee to provide basic administrative management and an information service system for enterprise and related individuals. The park itself is able to perform administrative management, office informatization, communication scheduling, and security assurance.

Establishing and accelerating the construction of the e-government platform in the park and improving administrative services are effective measures to deepen reform, build a service-oriented government, and guarantee the delivery of the governmental project service. At present, the penetration rate of informatization hardware in government departments is relatively low – the issuance of government policies and orders keeps using paper as the communication medium;

sending and receiving, collection and summary of official documents have not yet been fully digitalized. There is an urgent need for the implementation of e-government in the logistics park.

9.6.2 Overall structure

By building the e- government system and using the information platform to integrate information resources of the park and its enterprise, mutual information exchange and work management channels are built. This creates overall information sharing advantages and the work management mechanism to promote information sharing, render one-stop administrative services, and improve the efficiency of processing enterprise applications. This is shown in Figure 9-8.

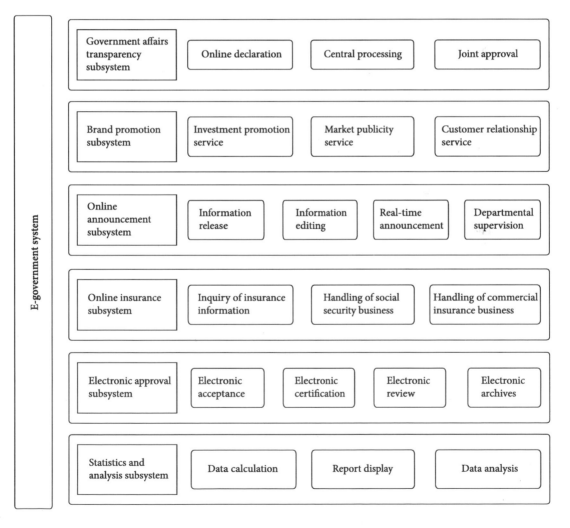

Figure 9-8 The overall structure of the E-government

1. Government affairs transparency subsystem

It discloses the work of the logistics park to the public as well as the tasks that it plans or is about to perform, such as park construction, road planning, public infrastructure measures, transaction processing, and their progress. Every enterprise can inquire and supervise this through specific channels, such as the government affairs transparency columns and networks.

The subsystem publishes all types of information. For example, new management systems can provide convenient and fast management services without employing a dedicated network company for maintenance. In a simple operation interface, the administrator can effectively manage news categories and content.

2. Brand promotion subsystem

Brand promotion refers to a series of activities in which companies shape their own product images to earn wide acceptance among consumers; it aims to enhance brand awareness. The consistency principle of online brand promotion stipulates that the positioning, brand connotation, audience, and packing of the promoted brand should be consistent with visual communication. To measure the cost and actual effects of online promotion, the subsystem can mine sufficient data to support the evaluation and report of the promotion effect so as to maximize the ROI (return on investment) of online promotion.

It can display the image and work dynamics of enterprise in an organized and sorted manner. In the subsystem, the webmaster can add and modify necessary topic columns and manage themed articles without any knowledge in webpage making.

3. Online announcement subsystem

It can, as a major Internet channel for information release, notify users or visitors of important announcements in the logistics park. With this subsystem, users can easily build a centralized, networked, professional, and intelligent multimedia system that provides cutting-edge professional services such as information editing, transmission, publishing, and management.

4. Online insurance subsystem

Online insurance allows insurance companies or new online insurance intermediaries to provide customers with information about insurance products and services through the Internet, and engage in online insurance, underwriting, and other insurance businesses. It directly completes the sale and service of insurance products while banks transfer the premiums to insurance companies.

The subsystem enables insured units to make business declarations, business changes, information inquiries, and other social insurance services on the Internet; insured individuals can inquire about social insurance payments, accounts, and benefits on the Internet, thus greatly improving public service and management of the system.

5. Electronic approval subsystem

Abiding by the management philosophy of lawful, open, fair, just, and efficient operation, the logistics park adopts information technology to improve service functions and office administrative efficiency. It also establishes an electronic approval system in accordance with approval policies, which is responsible for policy consultation and centralized joint approval. It improves cross-departmental administrative approval and simplifies work procedures.

The system is an electronic approval solution that allows efficient implementation and approval of all activities and related processes, including electronic acceptance, receipt, certification, review, etc. The administrative approval units are the main body responsible while their relatively fixed contacts are subject to approval.

9.7 Public service system

9.7.1 Demand analysis

The public service platform is aimed at the public needs of user groups in the park. By organizing, integrating, and optimizing resources, it provides a channel to share infrastructure, equipment, and information in order to reduce repeated investment, improve resource utilization, and strengthen information sharing.

Building this platform is conducive to improving the standardization, professionalism and openness of public services. The successful operation of the platform will not only reduce park operating costs, but significantly improve operating efficiency – especially the cluster development of SMEs.

9.7.2 Overall structure

The public service system is an important carrier and realization approach of public service in logistics parks. It enables collaboration and communication between enterprise, talent, and park managers, thus playing a major role in promoting industrial development and improving park environment. It also contributes to optimized park resource configuration, specialized division and collaboration of labor, and the development, transfer, and application of common key technologies. It usually comprises of financial support, public talent, science and technological innovation, e-commerce, facilities and equipment, public business, and statistical and analysis services. This is shown in Figure 9-9.

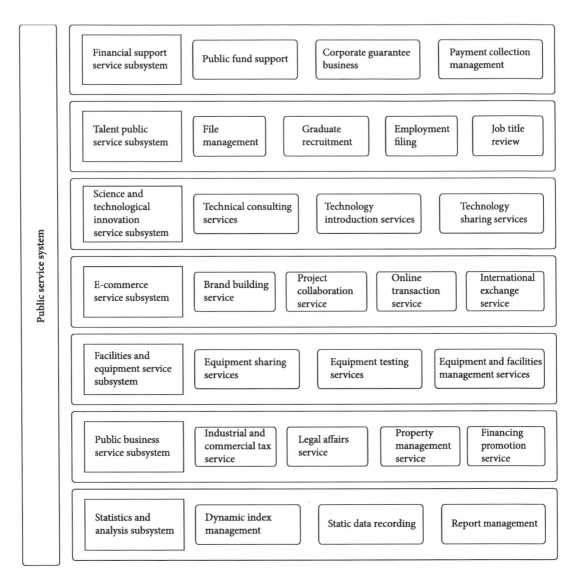

Figure 9-9   Overall structure of the public service system

1. Financial support service subsystem

Financial support means that logistics parks can focus on economic development and cultivating financial resources. It has continuously increased the support for park enterprise in terms of policies, public funds, and financial guarantee service, which has accelerated the pace of development of new products, science research, and technological transformation, and helped enterprise to grow stronger and transform and upgrade.

It provides public financial support services for enterprise, individuals and related entities, including financing solutions, re-guarantee business filing and re-guarantee fee collection, and government procurement-related product filing.

2. Talent public service subsystem

It adheres to the tenet of serving enterprise of all sizes, concentrates on improving employee quality of SMEs in the park and their capabilities, and through integration of high-quality social and park resources. It also positions itself as a supplier of talent quality improvement programs and ultimately strengthens the soft power of the park.

It renders public talent support service to enterprise, government departments and related institutions in the logistics park, including human resources consulting, file management, graduate recruitment, job title review, and employment filing, etc. There is also the talent introduction package service.

3. Science and technological innovation service subsystem

As a platform built to serve SMEs, promote industrial transformation and upgrading, and support enterprise innovation, it will focus on policy, investment and financing, technological innovation, talent, operation and management, platform, logistics park service, and credit service. This provides entrepreneurs with a one-stop service from seed incubation to achievement transformation, innovation, and development.

It usually renders public science and and technological innovation services to enterprise, government departments, related institutions, and individuals in the logistics park. With stronger resource integration and innovative resource sharing mechanisms, it will offer greater support for technological and enhance independent innovation capabilities.

4. E-commerce service subsystem

It aims to accomplish enterprise e-commerce activities, help enterprise manage production, sales, and service, support their external business cooperation, comprehensively improve informatization, and provide them with a computer network system with business intelligence. The subsystem relies on the public information network to enable external sharing and transmission of information and renders services regarding business needs, government affairs, and enterprise management of the park. This includes brand building, project collaboration, online transaction, international exchange, etc.

5. Facilities and equipment service subsystem

Public facilities and equipment are made available by the logistics park for the use of enterprise and employees in the park. By counting and integrating private equipment and facilities owned by enterprise, public facilities and equipment are allocated transparently and fairly – especially among enterprise in the same industry. Through registration, coordination, and remote operation, park resources are reasonably and effectively shared.

6. Public business service subsystem

Public business service is new. It involves basic public services (training and education, incubation support, etc.), economic public services (marketing and promotion, landmark establishment, etc.), social public services (accounting and auditing, industrial and commercial taxation, etc.), and public safety services (food hygiene and safety, network public safety, etc.). It is the fruit of Internet and science and technology development, and should be dealt with and constructed properly.

It provides public administrative services with the primary purpose of serving industry and enterprise development and improving management efficiency. Specifically, it covers industrial and commercial tax, legal affairs, property management, and financing promotion.

7. Statistics and analysis subsystem

It collects and sorts activities and data of the public service business, analyzes the data based on its development needs, feeds this back to government administrators and related enterprise in the form of reports and documents. It also helps users analyze the park's public service level, understand the needs of related enterprise and individuals, and improve service

9.8 Big data application service system

9.8.1 Demand analysis

As logistics park management becomes incrementally more complex, it will, as the core network chain, thoroughly alter the management model, business combination, business model, and operation mode of parks. Better tools must be used to quickly and efficiently explore the greatest data value.

In the big data era, combining the management and service status of parks, advanced technologies such as the IoT and cloud computing are adopted. They are aimed at platform applications to collect, store, and analyze data via data collection, data processing, data analysis, and mining, so as to extract useful information and provide intelligent and optimized decision support.

9.8.2 Overall structure

Based on the actual business needs of the park, it classfies, extracts, collects, integrates, and shares data from various business systems by promoting and widening the application of advanced information technologies. It is able to analyze and display business in a digital and graphic manner, provide management personnel with services, provide upstream and downstream enterprise and government departments with big data analysis services. This embraces six subsystems: data collection and storage, data combination, predictive analysis, smart business, operation analysis, and statistical analysis. This is shown in Figure 9-10.

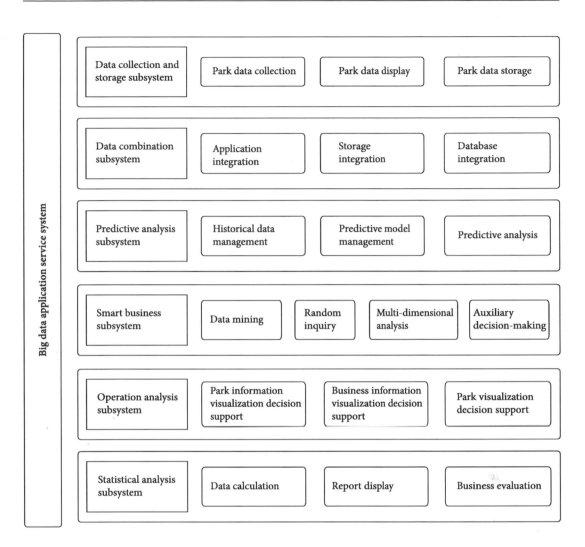

Figure 9-10   The overall structure of the big data application service system

1. Data collection and storage subsystem

As a major information technology, data collection of semaphore objects is carried out, analyzed, and filtered through processing mechanisms and stored. The comprehensive use of data collection, computers, sensors, and signal processing can establish a real-time automatic data collection and processing system. And data storage objects are temporary files generated by data streams during processing, or information to be searched during processing.

The subsystem collects and stores massive data generated in the commodity trading market, including commodity information, transaction, and customer data, etc.

2. Data combination subsystem

Data combination is an approach that collects, sorts, cleans, and converses data from different sources and loads it into a new source, providing consumers with a unified data view.

Considering individual needs of users, it provide a combination of services for product and transaction data to dynamically generate new big services, solving problems that a single big data service fails to meet. Meanwhile, it maximizes data value and improves its utilization.

3. Predictive analysis subsystem

Predictive analysis is a statistical or data mining solution that contains algorithms and techniques for use in structured and unstructured data to determine future results. It can be deployed for many other purposes such as prediction, optimization, forecasting, and simulation – providing planning information.

The subsystem combines modern forecasting methods and techniques to design and build a business forecasting system based on business data of relevant enterprise. It performs historical data management, forecasting model management, and forecasting analysis. It can combine data statistics and analysis results to provide an accurate and sufficient data basis to predict and judge business development trends.

4. Smart business subsystem

As a three-in-one system of computer Internet sites, mobile phone sites, and mobile WAP SMS group messaging, it applies the same domain name for access. In the unified management background of the website, the company's image websites are customized and enterprise related materials (company profile, product display, etc.) are added and written; the operation is simple and easy and the shared background of unified management synchronizes information release. Meanwhile, it can notify customers about the latest enterprise news through WAP SMS group messaging.

The subsystem, based on core, basic, and public business data of the logistics park, combines modern analysis methods and techniques to design business intelligence, which performs data mining, random inquiry, multi-dimensional analysis and auxiliary decision-making. It can provide comprehensive, multi-level decision support and knowledge services through automated index and report information management.

5. Operation analysis subsystem

Based on the data of other systems in the business operation support system, it builds a unified enterprise-level data warehouse. Advanced OLAP (online analytical processing) and data mining are employed to help the business decision-making layer understand the status quo of business operation, discover its pros and cons, and predict future trends. This is done to assist segment markets and customers, guide marketing and customer service departments to conduct targeted marketing and efficient customer relationship management, and to objectively assess the implementation and results of decision-making. This is widely accepted among users.

Based on basic information and sales information, it adopts GIS and other technologies to design and build the operation analysis subsystem, which is able to perform visualized decision support for parks and enterprise.

6. Statistical analysis subsystem

By employing data warehouse, multi-dimensional data analysis, and data mining technology, the financial systems are integrated and analyzed, thus useful information and knowledge are extracted to improve management and operation insights.

The subsystem combines relevant statistics, analysis and inquiry results through automated index and report information management to analyze business operation quality, risk, and management, and generate intelligent decision-making data. Consequently, it is able to provide relevant enterprise operators and managers with directional and multi-level decision-making support.

CHAPTER 10

RESEARCH ON KEY PLATFORM CONSTRUCTION AND CLOUD SERVICE TECHNOLOGIES

The stable operation of the smart information platform requires support from a key system. It is constructed by starting with the platform's planning goals and functional positioning, then key technologies are analyzed for each functional sector of the platform. The key technology system is standardized in a three-dimensional manner; it is composed of integrated system, data application, and information sharing and security technologies. This is shown in Figure 10-1.

10.1 Integrated system technology

Integrated system technology is a crucial safeguard for the comprehensive integration of various systems. It coordinates the operation and organic combination of separated subsystems to form a new comprehensive system to optimize performance. It includes information platform and system integration, software and hardware integration, and information sharing technologies.

10.1.1 System integration technology

This consists of system data integration, environmental support, business management and decision-making, standardization, and system development and implementation technologies. This creates permanent data, updated and searched for in various information systems; data access and integrity constraint rules can be set up and maintained. This is shown in Figure 10-2.

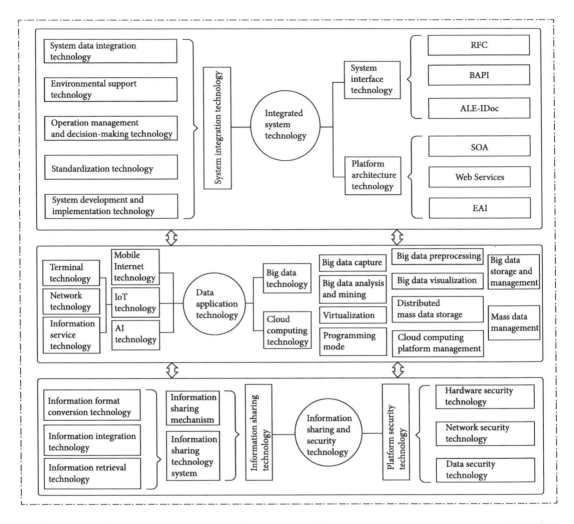

Figure 10-1 Key technology system of the smart logistics and supply chain information platform

1. System data integration technology

It includes data aggregation and concentration technologies. By applying this to various information systems, data access and integrity constraint rules can be set up and maintained.

1) Data aggregation and intermediary technology

It integrates multiple databases into a unified form. The data aggregate generated can be treated as a virtual database. Although it stores no data, it contains the content of multiple physical databases. Data aggregation and intermediary technology can both accomplish real-time data transfer and cross-platform management as required, and solve the problem of heterogeneous data platform integration.

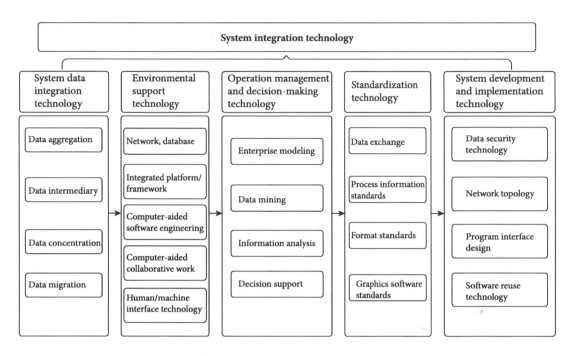

Figure 10-2   The structure of system integration technology

2) Data concentration technology

Involving data aggregation and migration, it logically or physically centralizes data of different sources, formats, and characteristics with conversion tools to enable comprehensive data sharing. There are three modes to achieve data concentration: federated, middleware, and data warehouse. The federated mode builds a data alliance, which is composed of multiple semi-autonomous database systems; the data sources provide mutual access and share data; while the middleware mode provides a unified data logical view in the middle layer to hide the underlying data details, thereby rendering a high-level retrieval service for heterogeneous data sources. The data warehouse mode builds a subject-oriented, integrated, time-related, and secure data collection, thereby providing data mining and decision support for the platform.

2. Environmental support technology

System integration requires solving all integration-oriented issues of hardware equipment and application software related to subsystems, architectural environment, and staffing; These all require environmental support. Environmental support technology includes networks, databases, and integration platforms/frameworks, computer-aided software engineering, computer-supported collaborative work, and human/machine interface technology.

3. Operation management and decision-making technology

The overall goal and strategic planning of the system involve many decisions. Operation management and decision-making technology breaks the limitations of information service, has an overall objective, and provides decision makers with multi-angle and multi-level decision-making support. Enterprise modeling, as a brand-new business management model, can provide enterprise with a framework structure to ensure their application system is closely matched with the business process of frequent improvement; system development and implementation technology paves the way for normal operation. Data mining extracts hidden, unknown, and useful information from a large spread of incomplete, noisy, and random data, thus supporting the decision-making of business managers.

4. Standardization technology

It involves data exchange, process information standards, format standards, and graphics software standards (also known as graphical interface standards). This refers to the interface standard for data transfer and communication on the system.

5. System development and implementation technology

Excellent system development and implementation technology can improve development efficiency, reduce repetitive work, and keep developed products from deviating from actual requirements. CASE (Computer Aided Software Engineering), which offers automatic approaches to document preparation and structural programming, is widely used in development, operation, maintenance, and management of computer software. It plays an supporting role in all stages of the development life cycle and saves time. The data of the logistics information system is not fully public: considering network security, data transmitted between subsystem interfaces must meet certain security and confidentiality measures to ensure safety. During system development, existing software elements are reused as much as possible so that development speed is accelerated and software productivity is improved.

10.1.2 System interface technology

Since the business content of modules on the smart platform varies, data sources are diversified across the entire system and data format. Therefore, when the system is integrated, system interfaces should be unified to ensure business modules are effectively connected. This connection, based on the interface, deploys corresponding management modules. Common system interface technologies are RFC (remote function call), BAPI (business Application programming interface), and ALE IDoc (Application Link Enabling Intermediate Document), which specializes in system integration. RFC is the foundation for both BAPI and ALE IDoc.

1. RFC

RFC is an essential and commonly used two-way interface technology between SAP (Systems Applications and Products) systems and other (SAP or non-SAP) systems. It is also regarded as the basic communication protocol between SAP and external parties. The RFC process involves a system calls program module outside the current system to achieve a certain function. This kind of remote function call can also be performed within the same system but usually they are in different systems.

2. BAPI

The BAPI interface is at the upper end of the RFC communication protocol. As a business object-oriented function interface built in SAP, it can directly process SAP business objects. When the external application system calls BAPI, it usually handles the SAP system's materials, orders, and other operational business entities as objects. This interface technology is frequently applied when mass data is imported and the external application system calls SAP.

3. ALE IDoc

ALE IDoc is the most widely and frequently used interface technology of SAP at present. Its biggest advantage is integrating heterogeneous platforms into a middleware technology layer unrelated to development platforms and development languages. Heterogeneous application platforms, between and within enterprise, all use this interface to interconnect, thus effectively solving data sharing and interaction problems.

10.1.3 Platform architecture technology

1. SOA

The overall design of SOA (Service Oriented Architecture) is service-oriented. SOA provides a new model of service-driven collaborative work for enterprise applications. It is a standards-based software development, organization, and design method that enables IT technology and applications to more closely serve business processes of enterprise, adapt to their dynamic changes, and shield technological heterogeneity and complexity via standardization.

Recently, many industries have launched their own service-oriented application platforms for enterprise IoT. The application of SOA can dynamically obtain and transmit data in a real-time manner, meaning a data service platform is available everywhere. This makes business processes more automated and intelligent and management more transparent.

2. Web Services

Web Services technology is a modular application that can be described, published, located, and allocated via the Web. As a common model to build applications, it can be implemented and run on all operating systems that support Internet communication. Web Services can perform all sorts of functions, from simple requests to complex business processes. Once it is deployed, other applications can discover and call the deployed service.

Applying Web Services technology can easily integrate heterogeneous systems of the platform without modifying the original system and affecting its functions. Adding a SOAP (Simple Object Access Protocol) interface to the original system interconnects the existing systems to allow mutual data exchange and access. In the railway-oriented information integration process, the original Web Services can be integrated through the system, which can also provide new Web Services to users. Within the railway transportation department, Web Services can be applied to RFID (radio frequency identification) systems.

3. EAI

EAI (Enterprise Application Integration) is a technique that integrates heterogeneous applications based on different systems with different solutions. By creating an underlying structure, it connects heterogeneous systems, applications, and data sources across the entire platform. This enables ERP (Enterprise Resource Planning), CRM (Customer Relationship Management), SCM (Supply Chain Management), database, data warehouse, and other significant data exchange methods. With EAI, departments can combine their core applications with new Internet solutions; it practically combines processes, software, standards, and hardware to achieve seamless integration between two or more platform systems.

10.2 Data application technology

10.2.1 Big data technology

Big data refers to a data collection that cannot be captured, managed, and processed with conventional software tools (within a certain time frame). It is an enormous, rapidly growing, and diverse information asset that exerts stronger decision-making power, insight, and process optimization capabilities. Big data is characterized by 5Vs: volume (great volume), velocity (high velocity), variety (great variety), value (low value density), and veracity (high veracity). Big data technology is used to quickly obtain valuable information from all types of data. Its key components are big data capture, preprocessing, storage and management, analysis and mining, and visualization. This is shown in Figure 10-3.

1. Big data capture

It obtains massive valuable data of various types via continuously developed collection techniques. The most common data types are texts, photos, videos as well as XML

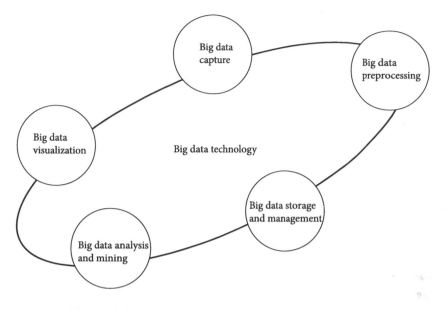

Figure 10-3   The key components of big data technology

2. Big data preprocessing

It identifies, extracts, and cleans the received data. It obtains valuable data through identification and analysis, and transforms complex data into a single or easy-to-process configuration, thereby accomplishing rapid analysis and processing.

3. Big data storage and management

It stores the collected data in memory, establishes a corresponding database according to specific business requirements, extracts, operates and analyzes the data, forms the target data required by enterprise, and manages and allocates it.

4. Big data analysis and mining

It extracts hidden and unknown but potentially useful information from a large amount of incomplete, noisy, ambiguous, and random practical application data.

5. Big data visualization

It analyzes big data to get certain patterns, selects and creates certain visualization methods based on observations of these patterns, and maps their numerical relationships through graphic or chromaticity space; thereby visualizing the data.

10.2.2 Cloud computing technology

Cloud computing is a form of data-intensive supercomputing. It enables usable, convenient, and on-demand network access to a configurable computing resource sharing pool (the resources are networks, servers, storage, application software, and services). Its key components are virtualization, distributed mass data storage, mass data management, programming mode, and cloud computing platform management. This is shown in Figure 10-4.

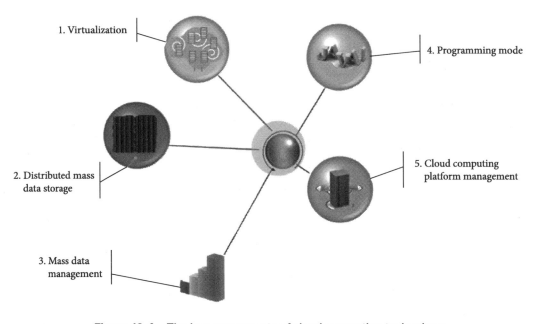

1. Virtualization

4. Programming mode

2. Distributed mass data storage

5. Cloud computing platform management

3. Mass data management

Figure 10-4   The key components of cloud computing technology

1. Virtualization
Computing elements run on virtuality instead of reality. They can expand hardware capacity, simplify software reconfiguration, reduce virtual machine-related overhead, and support a wider range of operating systems. Virtualization also isolates software applications from the underlying hardware (including a split mode that divides a single resource into multiple virtual resources, and an aggregation mode that integrates multiple resources into one virtual resource). Virtualization is widely used in CPU, operating systems, and servers; it is the best solution to improve service efficiency.

2. Distributed mass data storage
The cloud computing system serves a large number of users simultaneously. Therefore, it adopts distributed storage to store data, and redundant storage to ensure data reliability – this can guarantee high availability, high reliability, and economic efficiency of distributed data.

3. Mass data management

Cloud computing requires processing and analyzing large amounts of distributed data. The data management in the cloud computing system includes Google's Chubby data management and Hbase – the open source data management module developed by team Hadoop. How to find specific data and how to ensure data security and efficiency are problems that cloud computing data management must solve.

4. Programming mode

Cloud computing, as data-intensive supercomputing, is behind large-scale clusters and massive data, and objectively requires a distributed programming mode. In a cloud computing scenario, the programming mode must be able to analyze and process massive data easily and quickly, and provide security, fault tolerance, load balancing, high concurrency, and scalability, etc. Cloud· computing adopts a concise distributed parallel programming model – Map Reduce. This is a programming and task scheduling model mainly used for parallel operation of data sets and scheduling of parallel tasks.

5. Cloud computing platform management

It enables a large number of servers to collaborate, facilitate business deployment and activation, promptly discover and restore system failures, and achieve reliable operation of large-scale systems via automated and intelligent means.

10.2.3 Internet of things (IoT) technology

IoT embraces these key technologies: terminal, network, and information service: terminal technology is used to perceive "things," network technology to transmit and exchange "things," and information service technology to offer users with various information services. This is shown in Figure 10-5.

1. Terminal technology

It is used to perceive "things," including RFID, sensors, and embedded systems. The purpose of RFID is to identify objects and give each of them an "ID card"; sensors aim to timely and accurately obtain various information about external things, such as temperature, humidity, etc.; embedded systems strive to control, supervise, or manage the devices.

2. Network technology

It is used to transmit and exchange "things" related to information and services, including EPC (Electronic Product Coding), EPC ONS (Object Name Service), information service exchange, and wireless sensor networks. Among them, EPC aims at RFID to deliver and exchange item information services, thereby realizing automatic tracking management of the logistics supply chain. Information service exchange is based on EPC, oriented at all information services,

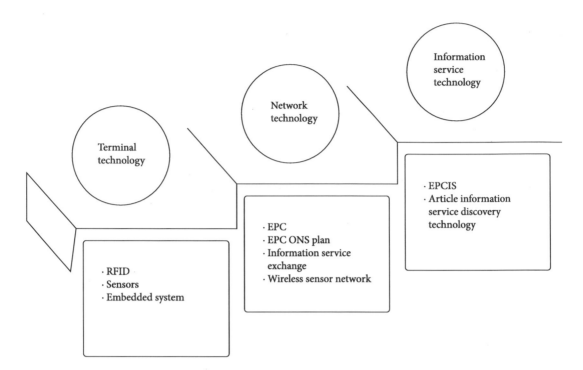

Figure 10-5   Key components of IoT technology

and integrates and shares information services. Wireless sensor networks employ wireless communication methods to enable information interaction between things and people.

3. Information service technology

The role of EPCIS (EPC Information Services) is the data storage center of the EPC Network; all data related is stored there. In addition to storage, it also provides a standard interface for information sharing. In the EPC Network, companies in the supply chain, including manufacturers, distributors, and retailers, are required to provide EPCIS, but their shared information varies. EPCIS adopts WebService to allow other applications or trading partners to inquire or update information through the interfaces. Only with EPCIS can we master specific product circulation processes and other product-related information.

10.2.4 Mobile internet technology

This is the product of integration and development of Internet technology and mobile communication. It uses mobile networks as the access network for mobile terminals and application services. The application network is the core of mobile internet and the ultimate goal of users. It is characterized by the ability to connect mutually dependent multiple heterogeneous groups to create value, realize information docking between enterprise and users anytime and anywhere, and bring a new platform-centered system oriented at user demand. In this case, users are not only the demand for innovation, but the supply of innovative resource data too.

1. Mobile terminal

Mobile terminals are the premise and foundation of mobile Internet technology. As mobile terminal technology keeps evolving, mobile terminals have gradually become equipped with stronger computing, storage, and processing capabilities as well as functional components such as touch screens, positioning, video cameras, and smart operating systems. Mobile terminal research covers not only terminal hardware, operating systems, software platforms, and application software, but also cutting-edge technologies. Among them, energy saving and positioning are paramount: better energy efficiency means greater endurance of mobile terminals, and obtaining terminal location is a prerequisite for using location-based services.

2. Access network

It is one of the essential infrastructures of mobile Internet. According to different network coverage, there are five main types of radio access networks: satellite communication, cellular (2G and 3G networks, etc.), WMAN (wireless metropolitan area network), WLAN (wireless local area network), and bluetooth-based wireless personal areas. Access network involve basic theories and key technologies of wireless communication and networks, including information theory and coding, signal processing, broadband wireless transmission theory, multiple access technology, heterogeneous wireless network integration, mobility management, wireless resource management, etc.

3. Application services

They are the core of mobile Internet. Distinct from traditional internet services, they are characterized by mobility and personalization. Users can receive mobile Internet application services anytime and anywhere; the services are customized based on user location, needs, and environment. Specifically, they are mobile search, mobile e-commerce, mobile Internet application expansion, cloud computing-based services, applications based on smartphone perception, etc. The applications of mobile Internet technology in logistics are usually positioning technology, RFID, and LBS (location-based services).

1) Positioning

At present, widely used positioning technologies are GPS, GIS, and positioning technology based on GSM (Global System for Mobile Communication).

GPS depends on navigation satellites to monitor distance and time. Any spot on earth can be located via their calculation. Currently, the most common vehicle positioning relies on GPS to obtain vehicle real-time location. Location information is sent to the logistics information control platform via mobile Internet and is processed and input to API (Application Programming Interface), which displays the real-time vehicle position, calculates arrival time, and generates transportation trajectory.

As an essential tool for acquiring, sorting, analyzing, and managing geospatial data, GIS is a a computer technology system established for geographic research and decision-making services. It collects, operates, analyzes, simulates, and displays spatial data, and adopts geospatial model analysis methods to promptly provide a variety of spatial and dynamic geographic information. The application of GIS promotes the sharing of logistics information, visualizes the logistics process, and reduces circulation costs.

The GSM-based positioning technology adopts the positioning principle of triangulation to calculate the specific position of mobile terminals, providing interaction between the communication base stations and the mobile terminal connected to them. Compared with GPS, it has higher positioning accuracy and stronger environmental adaptability, for instance, high-precision positioning still works in places where satellite information is poor like tunnels; and GPS can be combined to reduce blind spots.

Since positioning technologies promptly display the specific location of vehicles for the shippers, if any abnormality is found during transportation, it can be alerted in time, meaning greater safety of cargo transportation. Finally, they can be used to supervise the vehicle speed to improve safety.

2) RFID

It is an automatic identification technology that employs wireless radio frequency to perform non-contact two-way data transmission between the reader and the radio frequency card. The technology's long identification distance and high reliability free itself from the influence from all kinds of harsh environments. One of the rudimentary tasks in the digitalization of warehousing and logistics is the collection of basic data, whose authenticity and completeness are key to this digitalization. The application of RFID has greatly improved the efficiency and authenticity of data collection.

3) LBS

This is an information service that obtains mobile terminal location information by means of communication technology, determining the position of mobile targets and feeding it back to the mobile terminal. Modern positioning technologies are relatively abundant; they are being widely used on smart phone terminals, which receive information and data sent by satellites with known locations. They determine real-time location of their GPS receiver, calculate latitude and longitude, and display the corresponding location on the electronic map through module processing. When the present consumption pattern is gradually transformed to target consumption, LBS is the basis for realization of precision marketing in the future. They can run a comprehensive analysis based on customer location, the service provided by the surrounding environment, and the study findings of customer behavior to give personalized recommendations and maximize consumption volume.

10.2.5 Artificial intelligence (AI) technology

It is a technology that uses machines to simulate, materialize, and extend human intelligence. It draws on the idea of bionics and abstracts knowledge in mathematical language to imitate biological systems and human intelligence mechanisms. The technology can enable the logistics system to imitate human intelligence, to think, to perceive, to learn, to reason and judge, and to solve certain problems by itself. Regarding logistics, the application of AI focuses on smart search, reasoned planning, pattern recognition, computer vision, and intelligent robots.

1. Smart search

As a new generation of search engine that combines AI, it allows traditional quick search, relevance ranking, and intelligent information filtering and notification according to customer needs. For the smart platforms to apply smart search, it must first collect all basic logistics data promptly, and on the basis of fully stocked logistics data, intelligently analyze the true intention of customers who inquire about logistics services. This improves relevance and accuracy of the results returned by the search engine and personalizes smart searches based on their intention.

2. Reasoning planning

The application of AI to reasoning planning refers to the analysis of massive historical data, learning and summarizing corresponding knowledge from it, and building models to predict future data. In warehousing, to select the best warehouse location, AI can fully optimize and learn according to the constraints of the actual environment in order to propose a near-optimal selection plan. It can prevent the intervention of human factors, make site selection more accurate, save corporate costs, and increase corporate profits. In terms of inventory management, it dynamically adjusts inventory levels by analyzing historical consumption data and maintains the orderly circulation of corporate inventories. While improving consumer satisfaction, it avoids waste from blind production and ensures high-quality production. About distribution, the smart computing system automatically calculates possible delivery paths based on received orders and recommends the best delivery site and time; for precise delivery, it will not only accurately locate the customer's shipping address, but also recommends the most reasonable number of couriers and routes based on factors such as delivery time and efficiency – this optimizes the pairing of packages and couriers.

3. Pattern recognition

It refers to the use of mathematical knowledge and techniques to study the automatic processing and interpretation of patterns through computers. The application of AI in pattern recognition enables three-dimensional facial recognition, a biometric technology for identity recognition based on human facial feature information. Images or video streams with human faces are collected through cameras, and automatically detected in the images. With the wide application of smart express lockers in end distribution, the pickup methods welcome facial pickup in addition to manual inputing of pickup codes and scanning QR codes with mobile phones. The introduction of facial recognition has greatly improved the convenience and safety of pickup.

4. Computer vision

It refers to the use of machine vision, cameras and computers in lieu of human eyes to identify, track, and measure targets, and perform further graphic processing. Computer processing generates images more suitable for human eyes: in recent years, as computer vision evolves, especially video information detection and recognition technology, effective algorithms are built to associate low-level image processing technology with high-level video content analysis, thus promoting the application of computer vision in logistics. It can guarantee, in material circulation, there is no accidental transfer of materials due to information errors. Meanwhile, it is necessary to establish a complete information database for material transfer personnel to update and monitor the transportation status of each link promptly. This ensures that materials can be displayed with the fastest and most accurate information during a transfer. For example, when items are out of the warehouse and transferred to a certain site, users can track them in real time through online inquiries.

5. Intelligent robots

They are the intelligent tools that can replace traditional labor and form a powerful virtual labor force with productivity much higher than humans. In terms of end distribution, the research and development of drones and unmanned delivery vehicles solve the problem of last-mile delivery. In warehousing operation, AGVs (automated guided vehicle) can automatically identify the picking area and send the items to the designated shelves according to the planned routes. Cache robots can quickly collect the goods required in order and deliver it to the picking robot; finally, seeding robots on both sides of the assembly line load it into the courier box through a vacuum cup. From the user's order to delivery, the entire process is mechanized and efficiency improved.

10.3 Information sharing and security technology

10.3.1 Information sharing technology

It refers to the technology that enables exchange and sharing of information and products at different levels, and between different departmental systems, on the basis of information standardization and in accordance with laws and regulations. It aims to share information resources, optimize resource allocation, save social costs, improve utilization, and create more wealth. The information sharing technology of the smart platform consists of an information sharing mechanism and technology system.

1. Information sharing mechanism

The rudimentary concept of smart platform information sharing is based on the core business of the four major platforms (smart logistics information platform; bulk commodity supply chain information service platform; bulk commodity electronic trading platform; and logistics park information service platform. Software and hardware technologies are adopted to effectively address information sharing and collaboration issues of the new logistics business system and processes. On the one hand, information sharing must guarantee the safety, efficient transmission, and sharing of logistics information; on the other hand, it must meet the collaborative needs of logistics information resources. Therefore, it primarily has to be secure to ensure smooth progress of the operation. Specifically, it includes a security mechanism and various integration mechanisms that achieve collaboration at the information, organizational, and business levels.

2. Information sharing technology system

1) Information format conversion technology

The smart platform involves many heterogeneous data sources, thus the necessity to convert shared data of different formats into one format for storage. This is based on metadata and performs related processing before sharing to other business systems.

2) Information integration technology

The smart platform involves data sources of various business systems, thus making higher requirements for data application. It is necessary to effectively integrate all types of information for unified processing, and gather the massive information in heterogeneous resources of different geographical distributions for high-performance sharing of information. It includes system information integration, service chain information management integration, information filtering, and information gathering.

3) Information retrieval technology

There is a great deal of shared information and services on the platform. A quick inquiry to obtain shared information requires the support of retrieval technology. The embedded retrieval model and retrieval algorithm of the smart platform can speed up information and service sharing.

10.3.2 Platform security technology

Platform security refers to the timely delivery of correct information to those who have the authority to accept it, while preventing unauthorized users from illegally accessing and interfering with network resources. It involves hardware, network, and data security technology.

1. Hardware security technology

The security of the platform hardware has a direct impact on the development of platform business functions. Hardware-based security protocols can solve problems such as trusted computing based on third parties. The hardware security technology applied on the smart platform includes built-in security confirmation, external auxiliary security detection, and hardware development security management.

2. Network security technology

This is the security technology applied in information interaction to prevent unsafe attacks during the exchange and transmission of network information and data of IoT. As IoT information transmission relies more on wireless communication technology, network security technology includes VPNs, network antivirus, and network firewalls.

3. Data security technology

In the era of big data, information security is faced with a new normal. Throughout the big data life cycle of data collection, transmission, storage, management, analysis, release, use, and destruction, data is challenged by security threats. To ensure data security of the smart platform, plenty of technological security measures are taken, including collection, data transmission, data storage, data mining, and private data protection.

10.4 Cloud service design of the smart logistics and supply chain information platform

10.4.1 Cloud service of the smart logistics platform

1. Cloud service provider selection

In the era of big data, cloud computing has become the infrastructure for economic and social development. According to companies' need for informatization, the logistics industry cooperates with cloud service providers such as Alibaba and Baidu to build smart platforms. This stimulates efficient, virtualized, and generalized platform operations.

1) Alibaba cloud service

Alibaba Cloud is a public and open cloud computing service platform created by Alibaba Cloud Computing – a wholly-owned subsidiary of Alibaba. The data centers of Alibaba Cloud are located in Hangzhou, Shanghai, and Hong Kong. It has the largest CDN (content delivery network) in China, with more than 200 CDN nodes all over the country. The company's products are dedicated to improving operation and maintenance efficiency, reducing IT costs, and allowing users to focus more on core business development. Services rendered by Alibaba Cloud includes the underlying technology platform, elastic computing, database, storage and CDN, network load balancing, large-scale computing, cloud shield, and management and control, etc.

The core of the platform is the Apsara Cloud that Alibaba Cloud independently developed in February 2009. As a super-large-scale general computing operating system that serves the world, it connects millions of servers globally into a supercomputer to provide computing power in the form of online public services. Aspara's revolutionary character lies in the integration of the three directions of cloud computing: the provision of sufficiently strong computing power, general computing power, and inclusive computing power. The architecture of Apsara is shown in Figure 10-6.

Apsara Cloud is an intelligent mobile operating system that integrates cloud data storage, cloud computing services, and cloud operating systems. Based on Linux and WebKit (an open source browser engine), OpenGL and SQLite (lightweight databases), it offers underlying support of computing, storage and scheduling for upper-level Apsara services. This includes coordination, remote procedure calls, security management, resource management, and other common underlying services used to build distributed systems.

The Apsara open platform manages the physical resources of the Linux cluster in data centers, controls the operation of distributed programs, and hides the details of underlying failure recovery and data redundancy. It also links thousands of servers into a supercomputer and provides storage and computing resources to Internet users as public services. These interfaces and services include elastic computing and open storage, open structured data, and open data processing. Based on elastic computing, ACE (Ali Cloud Engine) is provided as a platform for third-party application development and web application operation and hosting.

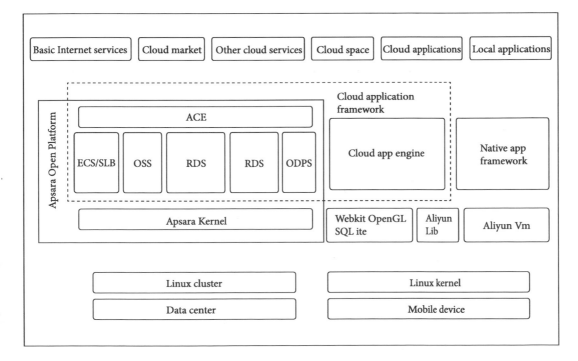

Figure 10-6 Apsara Cloud system architecture

2) Baidu Cloud service

Baidu Cloud is a public cloud platform created by Baidu. It has launched over 40 high-performance cloud computing products. The three smart platforms of Tiansuan, Tianxiang, and Tiangong, respectively, render smart big data, smart multimedia, and smart IoT services. Various industries can therefore access secure, high-performance, intelligent computing and data processing services; smart cloud computing becomes a new engine for social development. The service scope of the Baidu Cloud covers computing and networking, storage and CDN, databases, security and management, big data analysis, digital marketing cloud, intelligent multimedia, IoT, and AI. This is shown in Figure 10-7.

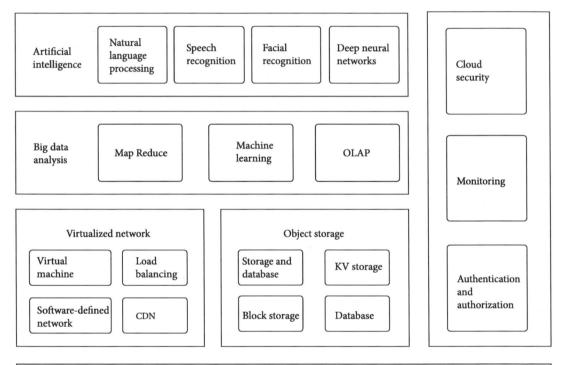

Figure 10-7   Baidu Cloud system architecture

Based on its highly reliable data centers, Baidu Cloud employs an advanced cluster management system to perform unified operation and maintenance management of servers, which greatly reduces the complexity of human maintenance and avoids human errors. Meanwhile, relying on intelligent scheduling technology, automate redundancy management is carried out upon deployed services to ensure their stability.

Baidu Cloud takes the lead in virtualization. With virtual machines and software-defined networking, multi-tenant isolation and cross-computer-room networking are realized. Customers are isolated from each other – invisible even if they are in the same computer room, which effectively ensures data security. Also, services deployed in multiple computer rooms can be incorporated into the same virtual network in a single region.

Baidu Cloud, with a variety of storing techniques, can give customers tailored solutions for different application scenarios. Whether it is a powerful and flexible database, a NoSQL (generally a non-relational database) storage system that pursues extreme performance, or a massive data backup at ultra low cost, it has a solution. All storage systems have gone through years of application practice in Baidu and have passed the large-scale pressure testing of massive data. Therefore, customer data is guaranteed to be safe and reliable.

Big data technology is a strong suit of Baidu. Baidu Cloud has different big data processing and analysis techniques such as Map Reduce, machine learning, and OLAP. Customers can extract information from the original log in batches and then use the machine learning platform for model training, or perform real-time multi-dimensional analysis of structured information. According to customer preference, different reports are generated to help business owners make decisions. Baidu Cloud can provide customers with complete big data solutions to maximize business data value.

It also has top-notch AI. The research findings of hundreds of top scientists are accessible to customers through Baidu Cloud. From text to audio messages to video clips, Baidu takes the lead globally. In deep learning, it is the most popular spot in the industry and is also at the forefront. Customers can utilize world-class AI through Baidu Cloud to make their business smarter.

2. Cloud service cooperation scheme

Based on the present IT assets and business needs of logistics companies, a hybrid cloud solution design can be adopted. This not only protects existing IT assets, but also promotes business development with the help from the service provider's cloud platform. Cloud services horizontally expand resources and seamlessly use big data, AI, and other open services to promptly build an independent and efficient business system. The hybrid cloud cooperation scheme of the smart platform is shown in Figure 10-8.

This scheme can reconnect the network between the private environment of logistics companies and the public environment of the service provider's cloud platform; synchronize and migrate public and private data, and exert big data processing capabilities to explore data value of supply chain and logistics operations; the interaction of public and private environments, and accelerate the construction of the four business application platforms.

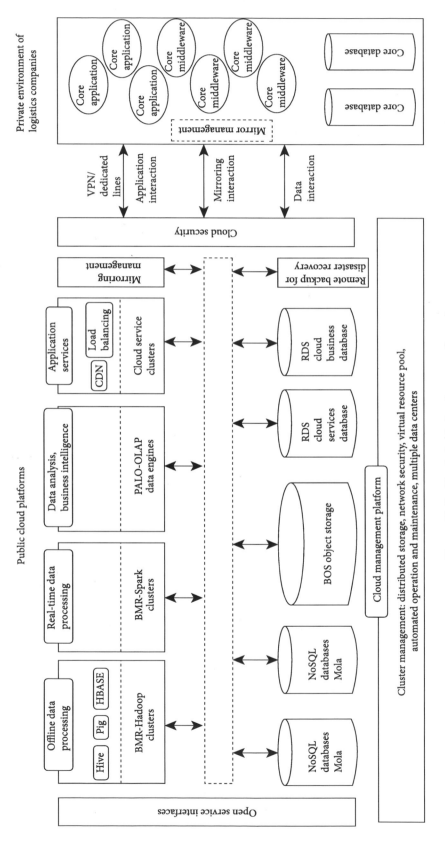

Figure 10-8 The hybrid cloud cooperation scheme

10.4.2 Cloud service model design

Based on the common needs of regional enterprise, logistics companies should fully exert resource, location, and industrial advantages to build a smart platform that conforms to the development trend of modern logistics. It should also comprehensively integrate existing regional public service resources to form an information service system covering the entire industry chain. Cloud computing technology can help the platforms build an IT basic information system, cloud data centers, basic networks, and cloud service system architecture. This is shown in Figure 10-9.

This cloud service architecture creates a corresponding virtual layer to the physical layer that covers servers, power systems, blade servers, storage devices, and network transmission equipment to provide cloud services, including IaaS (Infrastructure as a Service), PaaS (Platform as a Service), SaaS (Software as a Service) security, and operation and maintenance.

1. IaaS

IaaS refers to user utilization of all computing infrastructure, including CPU, memory, storage, network, and other basic computing resources. Users can deploy and run any software, including operating systems and applications. Corporate customers in the supply chain cannot manage or control any cloud computing infrastructure, but select operating system and apply storage space and deployment. IaaS includes the construction of optical fiber networks and data center computer rooms, the provision of basic IT resources, and the lease of supporting facilities to corporate customers and individual users in the supply chain.

2. PaaS

PaaS takes the software development platform as a service and submits it to users in the SaaS model. Corporate customers in the supply chain no longer need any programming to develop enterprise management software, including CRM (customer relationship management), OA (office automation), HR (human resources), SCM, and invoicing management – nor any use of other software development tools to launch it immediately online. It provides enterprise with a middleware platform for customized research and development while covering databases and application servers. The smart platform can cooperate with other domestic software developers to build and manage a software-level PaaS platform. Thus, corporate customers in the supply chain can quickly customize the workflow and database as they prefer and promptly build applicable enterprise management platforms.

3. SaaS

SaaS is a mode of providing software via the Internet. The smart platform deploys application software on its own server. Corporate customers in the supply chain can order the required application software services via the Internet as per their actual needs, pay according to the amount and duration of the subscription, and obtain the corresponding services through the Internet. They no longer need to purchase software, but instead rent web-based software from the information

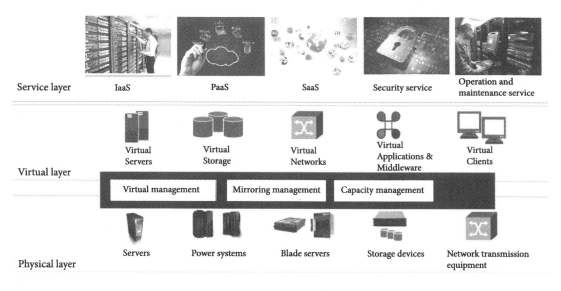

Figure 10-9 Cloud service architecture

platform to manage corporate business activities. This cloud service software includes enterprise email systems, OA systems, ERG (enterprise resource planning) systems, CRM systems, the HR system, the CAD (computer-aided design) system, and financial software etc.

4. Security
The smart platform can render various security value-added services, including security management, identity authentication and auditing, terminal authentication, and anti-spam, intrusion detection, to related enterprise while addressing the issue of internet access for supply chain corporate customers.

5. Operation and maintenance service
The smart platform can render 24/7 operation and maintenance services to corporate customers in the supply chain and undertake the daily maintenance of the company's internal information platform. Therefore, companies do not have to set up IT departments and professional technicians and so are able to reduce their investment and operating costs.

BIBLIOGRAPHY

[1] Bao, Xing, and Zheng Zhongliang. *Research on Supply Chain Financial Risk Control and Credit Evaluation.* Beijing: Tsinghua University Press, 2015.

[2] Cao, Jichang, and Duan Yunyong. "Research on Enterprise Business System Model and System Integration." *Computer Engineering and Science*, no. 2 (2003): 86–90.

[3] Chen Jin et al., *Financial Services for Bulk Commodity Trading.* Beijing: Chemical Industry Press, 2015.

[4] Chen, Hua, Ma Fengmei and Han Yanyan. "Research on XML-based Heterogeneous Data Integration Mode." *Microelectronics and Computer* 26, no.1 (2009): 137–139, 144.

[5] Chen, Shan. "Research on the Construction of Virtual Logistics Information Service Platform under E-commerce Environment." *Modern Trade*, no. 11 (2011): 232–233.

[6] Chen, Wenjun, and Xie Jiancang. "Building a Virtual Data Center to Integrate Flood Control and Drought Relief Information Resources." *Shaanxi Water Resources*, no. 4 (2006): 16–18.

[7] Chen, Xiaoliang, and Wang Chulin. "Extension of Smart Logistics from the Platform to the Supply Chain." *Internet Economy*, no. 11 (2017): 20–25.

[8] "China Federation of Logistics and Purchasing Interpretation Guiding Opinions of the General Office of the State Council on Actively Promoting Supply Chain Innovation and Application." *China Logistics and Purchasing*, no. 21 (2017): 30–32.

[9] Cui, Ruihua, Sun Jiaming and Meng Qinglong. "Into the Modern Technology 7: The Development Status and Trend of CIMS." *Journal of Electrical Engineering*, no. 10 (2008): 43–45.

[10] Deng, Xiaoyu, and Ma Weimin. "Research on the Content of Modern Logistics Industry Informatization under the Integration of Industrialization and Informatization." *Development Research*, no. 1 (2011): 52–55.

[11] Dou, Xin. "Comprehensive Information Service Platform Design for Logistics Parks with Cloud Computing and Internet of Things." *Modern Electronic Technology*, no. 11 (2017): 25–28.

[12] Fan, Chun. "Planning and Design of Regional Smart Logistics Information." *Electronic Government Affairs*, no. 7 (2012): 96–105.

[13] Fang, Meiqi. *Introduction to E-commerce*. 2nd ed. Beijing: Tsinghua University Press, 2002.

[14] Feng, Liang. "Regional Logistics Informatization Strategy and Implementation Research." *Logistics Technology* 34, no. 9 (2015): 72–74.

[15] Gong, Guan. "The Perspective of Information Technology – Building a Public Information Platform for Smart Logistics, Opening a New Era of Smart Logistics." *Logistics Technology*, no. 18 (2013): 87–90.

[16] Gu, Dongming. "The Construction of a Smart Information Service System Centered on Small and Medium Manufacturers." *Communication World*, no. 17 (2016): 249.

[17] Gu, Huiling, Zhu Jing and Li Lijie. "The Organic Combination of Supply Chain, Value Chain and Activity-based Costing." *China Market*, no. 32 (2007): 22–23.

[18] Guo, Jianbo. "A Preliminary Study on the Construction of a Cross-border Commodity Electronic Trading Platform – With Guangxi and ASEAN as an Example." *Northern Economy and Trade*, no. 11 (2017): 18–20.

[19] Guo, Yixing. "Discussion on the Construction of Enterprise Service Support Cloud for Smart Parks." *Telegraph Express*, no. 11 (2014): 11–13, 17.

[20] Han, Youtao, Ma Chun, and Zhao Xichang. "Association Rule Update and Algorithm." *Heilongjiang Science and Technology Information*, no. 18 (2007): 89.

[21] He, Miaomiao. "Research on the Information System Planning of Shenyang Offshore Logistics Parks." Beijing: Beijing Jiaotong University, 2007.

[22] He, Xin. "Research on the Construction of Auto Parts Industry Cluster and Logistics Public Information Platform." Wuhan: Wuhan University of Science and Technology, 2010.

[23] Hu, Lan. "When Big Data Meets 'Smart Park' – Suzhou Industrial Park Explores the Practice of Government Public Service Platform under the Cloud Mode." *China High-tech Zone*, no. 2 (2016): 72–75.

[24] Huang, Gang. "Research on the Application of Mobile Internet to Construct Public Service Systems." *Science and Technology Innovation and Application*, no. 20 (2013): 77.

[25] Jiang, Guohua, Li Xiaolin, and Ji Yingzhen. "Research on SOA-based Framework Model." *Computer and Information Technology*, no. 6 (2007): 37–39.

[26] Jiang, Min, and Xu Yan. "Application of Data Mining in University Teaching Management." *Computer Knowledge and Technology* 8, no. 24 (2012): 5741–5745, 5760.

[27] Jiang, Yihui. "Review of the Chinese Logistics Financial Development Theory Research." *Commodity Storage, Transportation and Maintenance*, no. 4 (2007): 7–9.

[28] Ke, Haisheng. "An Analysis of Logistics Financial Service Innovation from the Perspective of Sustainable Development." *National Management Informatization* 20, no. 1 (2017): 127–129.

[29] Li, Gang, and Wei Quan. "Enterprise Information System Integration Framework Based on the Synergy Theory." *Journal of Library Science in China*, no. 6 (2006): 61–64.

[30] Li, Qin. "Discussion on the Construction of Regional Logistics Information Platform." *Computer Knowledge and Technology* 13, no. 21 (2017): 247–248.

[31] Li, Xiangwen. *Digital Logistics and E-logistics*. Beijing: China Materials Publishing House, 2011.

[32] Li, Xin, and Liu Fengmei. "Smart Logistics Information Support Technology Based on Big Data." *Southern Agricultural Machinery* 48, no. 18 (2017): 20.

[33] Li, Yixue, Wang Shouyang and Feng Gengzhong. *Logistics and Supply Chain Finance Review*. Beijing: Science Press, 2010.

[34] Lin, Jun. "Baicheng Power Plant Management Information System Design and Application." Changchun: Jilin University, 2015.

[35] Liu, Hangyuan. "Research on the Information Platform Construction of Smart Logistics Parks." *Information Technology and Informatization*, no. 9 (2016): 123–125.

[36] Liu, Qiong, Cui Shouling and Ye Jingjing. "Research on the Fourth Party Logistics Service Platform Based on SOA." *Mechanical Design and Manufacturing*, no. 9 (2007): 210–212.

[37] Liu, Shanshan. "Reconstruction of Supply Chain Value in Smart Logistics." *File*, no. 34 (2017): 233.

[38] Lou, Qianfei. "Discussion about the Linkage between E-commerce and Logistics." *Economic Forum*, no. 12 (2006): 69–71.

[39] Lu, Mingzhu. "Enterprise Logistics Informationization Strategic Planning and Implementation Research." *Logistics Technology* 33, no. 21 (2014): 438–440.

[40] Luo, Renshu. "The Construction of a Smart Logistics Information Platform." *Logistics Engineering and Management*, no. 1 (2014): 80–81.

[41] Luo, Ya. "The Upgrade Project of Kunshan Small and Medium-sized Enterprise Information Platform." *Jiangsu Science and Technology Information*, no. 8 (2009): 43–45.

[42] Mao, Yu, and Dai Yuehong. "Service-based Enterprise Application Integration." *Science Technology and Industry*, no. 1 (2007): 38–41.

[43] Meng, Gaoyuan. "Research on the Framework of Integrated Logistics Information Platform and Development of Land Transportation Business Transaction and Transportation Scheduling System." Nanning: Guangxi University, 2013.

[44] Niu, Luqing. "Alibaba Cloud: Innovative Cloud Computing." *New Economic Guide*, no. 3 (2013): 66–68.

[45] Ouyang, Xiaobin. "Discussion on the Integration of Library Information Systems." *Journal of Zhengzhou University*. Philosophy and Social Sciences Edition, no. 6 (2002): 142–145.

[46] Shen, Zemin. "The Design and Realization of Modern Logistics Transportation Management System." *Logistics Engineering and Management* 39, no. 5 (2017): 80–81.

[47] Sheng, Yuhua, and Pan Chichun. *Supply Chain Management and Virtual Industry Chain*. Beijing: Science Publishing House, 2004.

[48] Shi, Rongli. "Information Platform Construction of Smart Logistics Parks Based on Big Data." *Enterprise Economy*, no. 3 (2016): 134–138.

[49] Song, Hua, and Hu Zuohao. *Modern Logistics and Supply Chain Management*. Beijing: Economic Management Press, 2000.

[50] Su, Ni. "Discussion on Logistics Financial Value-added Services." *Business Culture*, Academic Edition, no. 1 (2008): 98.

[51] Sun Shaoling et al., "Research and Implementation of Cloud Computing and its Application." *Telecommunications Engineering Technology and Standardization* 22, no. 11 (2009): 2–7.

[52] Sun, Bin, and Wang Dong. "Big Data System Construction of Belt Road Initiative Logistics Centers." *China Circulation Economy*, no. 8 (2017): 32–40.

[53] Tang, Hui. *An Analysis of the Standard System of Logistics Public Information Platform*. Beijing: Publishing House of Electronics Industry, 2016.

[54] Tang, Shaolin. "Smart Logistics Park Information Platform Solution Based on Internet of Things Technology." *Logistics Technology*, no. 8 (2016): 70–73.

[55] "The Fourth National Logistics Park (Base) Investigation Report." *Logistics Technology and Application*, no. 11 (2015): 110–114.

[56] Wang Qi et al., "Application of 6 Sigma in System Integration." *System Engineering*, no. 12 (2006): 111–115.

[57] Wang, Hui. *Research on Third-party Logistics Companies Based on Supply Chain Management*. Beijing: Metallurgical Industry Press, 2015.

[58] Wang, Jian, and Yang Zhendong. "Discussion on Cloud Computing-based Application Mode of Small and Medium-sized Enterprises' Financial Information." *China Management Information Technology* 12, no. 17 (2009): 53–54.

[59] Wang, Ran. "Research on the Nature and Construction of Bulk Commodity Electronic Trading Platform." Zhoushan: Zhejiang Ocean University, 2013.

[60] Wang, Xifu, and Shen Xisheng. *Modern Logistics Informatization Technology*. Beijing: Beijing Jiaotong University Press, 2015.

[61] Wang, Xifu, Shen Jinsheng and Guan Wei. "Research on the System Framework of Regional Modern Logistics Public Information Platform." *Logistics Technology*, no. 1 (2006): 77–79, 83.

[62] Wang, Xifu. *Big Data and Smart Logistics*. Beijing: Tsinghua University Press, 2016.

[63] Wang, Xing, and Jiang Zhinong. "Application of XML-based Middleware Technology in Fault Diagnosis Data Integration." *Machinery Manufacturing and Automation*, no. 5 (2008): 112, 114, 122.

[64] Wang, Xinyu. "Logistics Integrated Information Management Platform Function and Architecture Design of Lanzhou Railway Administration." *Railway Transport and Economy*, no. 5 (2017): 54–58.

[65] Wang, Xinyue. "Research on Chinese Smart Logistics Development Problems and Countermeasures." *Railway Transportation and Economy*, no. 4 (2017): 37–41.

[66] Wang, Yafei, Shi Xinyi and Guan Zhichao. "Research on the Architecture of Service-oriented SOA Urban Intelligent Transportation Information Platform." Natural Science Edition, *Journal of Sun Yat-sen University* 49, no. S1 (2010): 23–28.

[67] Wei, Yuxin. "Application of Internet of Things Finance in the Electronic Trading Market of Bulk Commodities." *Shopping Mall Modernization*, no. 28 (2016): 35–36.

[68] Wu, Junqiang, Zhou Jiliu, and He Kun. "LLE-based Colored Facial Detection." *Journal of Sichuan University*, Natural Science Edition, no. 4 (2006): 796–800.

[69] Wu, Qinglie. *E-commerce Management*. Beijing: Mechanical Industry Press, 2009.

[70] Xia, Jia. "Analysis and Countermeasures of Bulk Commodity Electronic Trading Market." Chengdu: Southwest Jiaotong University, 2014.

[71] Xiao, Liang. "The Reorganization Model of Logistics Service System from the Perspective of Industrial Cluster Development." *Management Modernization*, no. 2 (2007): 12–14.

[72] Xie Xiaoxuan et al., "Research on the Integration of E-commerce and Enterprise Information Systems." *Computer Integrated Manufacturing System*, no. 5 (2002): 371–373.

[73] Xie, Hualong. "Modern Innovation and Application Research of Enterprise Logistics Management." *Modern Economic Information*, no. 2 (2015): 111–112.

[74] Xu, Jinli, and Deng Yanwei. "Research on the Construction of Coal Logistics Information Platform." *China Coal*, no. 2 (2013): 70–73, 80.

[75] Yang, Junai, and Zhang Yan. "An Analysis of the Development Model of Internet + Supply Chain Finance – With H platform as an Example." *Logistics Engineering and Management* 38, no. 7 (2016): 20–22, 32.

[76] Yang, Ling. "Design and Implementation of an Integrated Collaborative Office Platform for Group Enterprises." Shanghai: Shanghai Jiaotong University, 2009.

[77] Yang, Qingsong, Yang Xiaonong, and Yang Xuehai. "Research on the Supply Chain Financing Risks of the Bulk Commodity Spot Electronic Trading Market." *China Business*, no. 3 (2014): 98–100.

[78] Yang, Xiaoyu. "Discussion about the Application of Logistics Management Software in the Unit Information System." *China High-tech Enterprise*, no. 21 (2008): 44.

[79] Yang, Yang. "Beijing-Tianjin-Hebei Integration and Coordinated Development of Regional Smart Logistics." *Economic and Trade Practice*, no. 13 (2017): 133.

[80] Ye, Jianing. "The Construction of Logistics Information Platform." *China Road Transport*, no. 4 (2015): 78–79.

[81] Yue Jianwei et al., "Research on Collaborative Development Platform for E-government System." *Computer Engineering*, no. 3 (2007): 273, 274, 277.

[82] Zhang Shuhui et al., "Status Quo of Networked Manufacturing Platform and Technological Trend Analysis." *Computer Engineering and Applications*, no.5 (2006): 193–196, 200.

[83] Zhang, Guangbin. "Logistics Informationization and Industry Trend Analysis." *Value Engineering*, no. 31 (2017): 93–94.

[84] Zhang, Jianlai. "Basic Principles of RFC in SAP System and Its Application in Enterprises." *Shandong Industrial Technology*, no. 19 (2015): 294.

[85] Zhang, Jianxin. "An Analysis of Logistics Industry Clustering in Regional Economic Development." *Economic Issues*, no. 1 (2007): 119–121.

[86] Zhang, Jinxia. "Category Management Strategy and Case Study." Jinan: Shandong University, 2005.

[87] Zhang, Li, and Zhao Zheng. "Thoughts on the Construction of Logistics Information Platform in the Era of Big Data." *Shopping Mall Modernization*, no. 14 (2017): 71–72.

[88] Zhao, Xianglong. "Research on the Operation Mode of Chinese Bulk Commodity Trading Platform." Shanghai: Shanghai University of Engineering Science, 2015.

[89] Zhao, Yiqiang, and Zeng Junfang. "Web Services Application Review in RFID System." *Computer Application Research*, no. 12 (2006): 13–14.

[90] Zheng, Qingchang, Tan Wenhua and Huang Jinghan. "Research on the Construction of the Public Technical Service Platform for Industry in Fujian Province." *Journal of the Fujian Provincial Committee Party School of the Communist Party of China*, no. 1 (2008): 70–72.

[91] Zheng, Weifeng. "Design and Implementation of Civil Aviation Information Platform Services." Nanjing: Nanjing Aerospace University, 2008.

[92] Zhou, Binghai, Yang Zhibo and Xi Lifeng. "Standard Research on Networked Manufacturing and System Integration." *Computer Integrated Manufacturing System*, no.9 (2005): 1248–1254.

[93] Zhou, Kaile, Ding Shuai, and Hu Xiaojian. "A Summary of Research on the Internet of Things Information Service System for Massive Data Applications." *Computer Application Research* 29, no. 1 (2012): 8–11.

[94] Zhou, Yun, and Yin Lu. "Research on the Intelligent Development of Agricultural Products Supply Chain Supported by Modern Logistics System." *Agricultural Economics*, no. 9 (2017): 120–122.

[95] Zhu, Jie, Li Juntao, and Zhang Fangfeng. *Logistics Public Information Platform Construction and Operation Mode*. Beijing: Mechanical Industry Press, 2014.

[96] Zhu, Xiaoli. "The Format and Path of the Transformation from Information System Integration to Information Technology Services." *Electronic Technology and Software Engineering*, no. 21 (2017): 260.

[97] Zhu, Zhixian. "Research on the Development Problems and Countermeasures of Chinese Bulk Commodity E-commerce." *China's Collective Economy*, no. 30 (2017): 88–89.

ABOUT THE AUTHORS

Xifu Wang, Professor of Logistics Engineering at Beijing Jiaotong University, Traffic and Transportation school; PhD supervisor; smart logistics expert.

Professor Wang also takes part in some part-time work ventures. This includes, 'talents of Taishan industry in Shandong Province; senior consultant of Fujian's "One Belt And One Road" construction; member of the Expert Committee of CFLP Information Platform; Council member of China Transport Association; Evaluation expert of modern service fields at the Ministry of Science and Technology; Secretary General of the Editorial Board of Big Data and Smart Logistics Series, Ministry of Industry and Information; Expert member of CATIS; and Standing member of SESC.

Major academic contributions and achievements include: (1) Theory and application methods of big data and smart logistics; (2) Modular theory system of smart logistics; (3) Industry application-oriented intelligent logistics big data service platform; (4) Theory and application of supply-demand balance in the logistics industry; (5) Intelligent logistics business system and process; (6) Modern logistics platform based on IoT, etc.

Zhongfu Cui, born on December 4, 1961 in Xiantao, Hubei Province. Cui graduated from Renmin University of China with a major in industrial enterprise management in 1986.

Cui is currently the Vice President and Secretary-General of CFLP, and Chairman of Asia Pacific Logistics Alliance. His research interests are logistics management and integrated transportation.